Biology of Indian Morels

Biology of Indian Morels

T. N. Lakhanpal
Onkar Shad
Monika Rana

I.K. International Publishing House Pvt. Ltd.
NEW DELHI • BANGALORE

Published by
I.K. International Publishing House Pvt. Ltd.
S-25, Green Park Extension
Uphaar Cinema Market
New Delhi–110 016 (India)
E-mail: info@ikinternational.com

ISBN 978-93-80578-24-8

Published by Krishan Makhijani for I.K. International Publishing House Pvt. Ltd., S-25, Green Park Extension, Uphaar Cinema Market, New Delhi–110 016. Printed by Rekha Printers Pvt. Ltd., Okhla Industrial Area, Phase II, New Delhi–110 020.

Objects in nature have different forms, appearances and names. The truth, however, is that they have a common origin.

In this truth of creation, love is the energy that binds all the five elements together. The entire creation is suffused with the Divine energy of love.

To the Divine Lotus Feet of
Sri Sathya Sai Baba

&

To my guru and mentor
Shri K.G. Mukerjee

Foreword

I am happy to write a foreward for the monography 'Biology of Indian Morels' penned by Prof. T. N. Lakhanpal and his associates, Dr. O. S. Shad and Dr. Monika Rana. Morels have attracted the poor and the rich, the young and the old alike since times immemorial all over the world because of their taste and excellent flavour. Called 'Guchhi' in the Indian market morels are one of the most important commercial wild edible fungi. The research on the morels has been primarily centred on taxonomy and on the efforts for domestication. All the biological aspects, however, are perhaps being covered for the first time in this monograph.

Morels in India grow mainly in the Himalayan region that is why most of the research work has come from researchers located in and around the Himalayan region especially Jammu and Kashmir and Himachal Pradesh. Though different aspects have now been worked out, the artificial cultivation still eludes the researchers and that is the reason for so high cost presently. The authors of this monograph have investigated during the last three decades and have for the first time attempted to compile and consolidate the available literature along with their own work into this voluminous monograph.

There are fourteen chapters in this monograph each chapter is devoted to one aspect of morel biology. The monography begins with the overview provides information on life cycle pattern, given classical concept of morel taxonomy followed by molecular characterization of the species. Subsequently, it provides information on various ecological, physiological, rhizomorphospheric aspects and associations, cytology, mycological, sociobiological, nutraceutical potential and culinary aspects. Information collected from the local people and the impact of morel collection on the economy of the people has been discussed in detail. The data on medicinal uses, collection, drying and trading, nomenclatures, colour variation, relationship to snowfall and forest types reveals many startling facts about these small organisms with big potential. In nutshell this monograph provides excellent original information on almost all aspect of the morel biology. May be this compilation inspires researcher and one day the technology for its cultivation is developed and perfected. That will be the golden day in the history of the fungal biology and biodiversity. I congratulate Prof. T. N. Lakhanpal, Dr. O. S. Shad and Dr. Monika Rana for sincere efforts and hard work in preparing the Monograph 'Biology and Indian Morel' in a highly scientific and qualitative way and the monograph will be highly useful to Teachers, Scientists, Scholars and Students. I am sure the monograph will find ample use in all the libraries and research institutes.

Prof. Dr. S. Kannaiyan

M.Sc., (Ag.) Ph.D., C. Biol., M.Biol., (London)

FIBA (Cambridge), FABI (USA), FNAAS, FNABS, FISB, FAMI

Chairman

National Biodiversity Authority

Govt. of India

Preface

Morels have been 'moral' booster mushrooms as they have inspired mycologists to keep research on them alive so as to achieve their artificial cultivation. Morels are king mushrooms and have attracted human attention since time immemorial.

They have earned appreciation for their excellent flavour and culinary properties. Technically morels belong to the genus *Morchella* and are commonly called 'Guchhi' in the Indian market. They are presently the costliest mushrooms selling at Rs. 7000-8000 per kg in the Indian market.

The senior author was fascinated by the morels as a postgraduate student in the late sixties at the Botany Department, Punjab University, Chandigarh. He was exposed first to the colourful world of fungi when he joined fungal forays to the N-W Himalayas with groups led by late Prof. K.S. Thind. The interest was revived in the beginning of 1980s after he joined Department of Biosciences H P University Shimla in 1976. Systematic work on morels was initiated and executed from 1980 onwards with great zeal. This progressed further with the enthusiasm of the second author who worked for his M.Phil. (1985) and then Ph.D. (1989) on classical aspects followed by the third author who completed molecular characterization of the species for her Ph.D. in 2005. In between some interludes on ethnomycology were provided by Mangala Gowry, Meenakshi Rathod and Mitu Sharma.

Hence, this monograph on 'Biology of Indian Morels' is an outcome of studies initiated in 1980 and carried through all the years till 2005 steadily but continuously and earnestly. During all these years almost all aspects of morel biology have been investigated. However, the cultivation of morels still eludes the scientists the world over.

The monograph has a number of chapters, each devoted to one aspect of study. It begins with an overview of morel biology, the life cycle pattern, the classical and molecular systematic treatment, the ecological and physiological aspects, the nutritional and nutraceutical components, mycorrhizal and sociobiological and ethnomycological investigations, cytological studies and the traditional and modern recipes for morel consumptions. All the chapters are documented with tables, figures and original photographs.

It is hoped this monograph will interest botanists and in particular mycologists the world over. Suggestions for improvement are welcome.

T.N. Lakhanpal
Onkar Shad
Monika Rana

Acknowledgements

Many individuals, directly or indirectly, have helped in the accomplishment of this endeavour and mere acknowledgement of their help in a few formal words may not actually be truly representative of the assistance rendered. First and the foremost acknowledgement are to the Almighty for enabling us to undertake the task and successfully accomplishing it. Without His grace and unseen hand, nothing would have been possible. Our grateful thanks to Prof. K.S. Thind, Panjab University, Chandigarh, who inspired the senior author to undertake this type of work, to Prof. K.G. Mukerji. University of Dehli for the role of a friend, philosopher and guide in the execution and publishing of this work. Thanks in no less measure to Professors C.V. Subramanian, R.L. Munjal, J.N. Kapoor, T.N. Kaul, K. Natrajan, R.P. Purkayastha, Dr. A.V. Sathe, for help in various ways.

Thanks are also due to Dr. Roy Watling, Edinburgh, Dr. Gary Lincoff, U.S.A., Prof. James Trappe, Michael Castillano, David Arora, Randy Molina, Nancy Weber Smith, Oregon, Corvallis, U.S.A., Prof. E. Horak, Switzerland and Prof. T. Hongo, Japan. The senior author is thankful to all his students and colleagues who were ever ready for rendering help on all occasions.

Grateful thanks are also due to Prof. H.S. Garcha, PAU Ludhiana, Dr. R.P. Tiwari, Director, Dr. S.R. Sharma, NMRC, Solan, Prof. C.L. Jandik, Solan, Prof. Dalip Malhotra, Prof. S.K. Sood, Prof. V.K. Mattu, Dr. K.S. Kapoor, and Dr. Vineet Jistu for moral support and encouragement. Help extended by Dr. S.K. Chakrabarti CPRI Shimla during the studies on molecular aspect is gratefully acknowledged.

Our grateful thanks to our parents and family members for their understanding and encouragement.

Financial help from the Department of Science and Technology, and Dept. of Biotechnology, GOI, New Delhi to the senior author is duly acknowledged. This grant made possible the execution and the completion of the work.

T.N. Lakhanpal
Onkar Shad
Monika Rana

Contents

Abbreviations and Symbols

"	–	inches
%	–	per cent
/	–	per
&	–	and
µg	–	microgram
µl	–	microlitre
µm	–	micrometre
$	–	Dollar
AD	–	Anno Domini
amsl	–	above mean sea level
BC	–	Before Christ
BMR	–	Basic Metabolic Rate
bp	–	base pairs
c.f.	–	cited from
cms	–	centimetres
conc.	–	concentrated
d.w.	–	dry weight
Dia.	–	diameter
dil.	–	dilute
DNA	–	Deoxyribonucleic Acid

e.g.	–	for example
ed.	–	edition
Ed.	–	editor
Eds.	–	editors
et al.,	–	et. alia; and others
etc.	–	et cetera
Fig.	–	Figure
g	–	gram
H.P.	–	Himachal Pradesh
H_2SO_4	–	Sulphuric acid
HCL	–	Hydrocholoric acid
HDP	–	Health Defence Potentiators
HPUB	–	herbarium, Department of Biosciences, H.P. University, Shimla
hrs.	–	hours
i.e.	–	that is
ICBN	–	International Code of Botanical Nomenclature
J&K	–	Jammu and Kashmir
Kcal	–	Kilo calories
Kg	–	Kilogram
l	–	litre
l.c.	–	*Loco citato:* at the place cited
m	–	metre
mg	–	milligram
mgs	–	miligrams
min	–	munutes
ml	–	millilitre
msl	–	mean sea level
N.W.	–	north West
ng	–	nano gram
°C	–	Degree Celsius
PCR	–	Polymerase Chain Reaction
PDA	–	Potato Dextrose AGar
ppm	–	parts per millon
RAPD	–	Random Amplified Polymorphic DNA
RFLP	–	Restriction Fragment Length Polymorphism
rpm.	–	rate per minute
sp.	–	Speces (Singular)
spp.	–	Species (Plural)

sq km	–	Square kilometre
sq m	–	Square metre
Syn.	–	Synonym
TDF	–	Total dietary fibre
U.P	–	Uttar Pradesh
var.	–	variety
vern.	–	vernacular
viz.	–	videlicet; namely
vol.(s)	–	volume(s)
v/v	–	volume/volume
w.r.t	–	with respect to

CHAPTER 1

Biology of Morels–An Overview

CHAPTER OUTLINE

- Introduction
- Habit and Habitat
- Systematic Treatment
- Nutritive Value
- Ecological Aspects
- Rhizomorphosphere Associations
- Ethnomycological Significance
- Production and Trade

INTRODUCTION

The morels comprising the genus *Morchella*, are commonly referred to as 'Guchhi' in the Indian Market. At present, it is the costliest and most sought after edible fungus in the world. The morels are highly appreciated for their culinary aspects and gastronomical delights and excellent flavour. Even the mycelia are known to possess the same nutritive value as the sporophores (fruit body). The fructifications constitute the item of commerce and are exported from India to Europe and United States of America.

HABIT AND HABITAT

Species of *Morchella* in India occur primarily in N-W Himalaya, though there are some reports from sub-hilly areas and plains as well. The species grow on soil rich in organic matter, in loamy soil, in pure and mixed coniferous forests, broad leaved forests, in open areas or grasslands, in apple orchards or under various types of shrubs in the open under direct and indirect shade. Morels mainly appear during the spring season from March to May when snow begins to melt and is accompanied by some precipitation. Species of *Morchella* have, however, been recorded/collected during the rainy season as well (Wakode, 1983; Lakhanpal and Shad, 1986 a, b; Shad, 1989, Kamal, 2005 and Kaviyasaran *et al.*, 2006), in the months of September and October, repeatedly from some parts of India and various parts of Himachal Pradesh including lower hills which do not experience snowfall (Lakhanpal and Shad, 1986 a, b). *Morchella* fruiting bodies have also been reported from paddy field bunds (in Jammu) and maize fields (in Solan) (Jandaik and Sharma, 1995).

Majority of the *Morchella* species have been collected from N-W Himalayan region, especially Jammu & Kashmir (Cooke, 1870; Sydow and Butler, 1911; Ghosh and Pathak 1962; Batra and Batra, 1963; Waraitch, 1976 and Kaul, 1976, 1981), Himachal Pradesh; (Sohi *et al.*, 1965; Lakhanpal and Shad, 1986) and Uttar Pradesh (Hennings, 1901 and Theissen, 1911).

Occasional reports about collection of *Morchella vulgaris* Boud. from Assam forests, *M. deliciosa* from Amritsar (Punjab) and *M. conica* Pers. from central India have also been given by Bhattacharya and Baruah (1953), Purkayastha and Chandra (1985), and Ghurde and Wakode (1981), respectively. The validity of *M. vulgaris* from Assam forests, however, has been doubted by Ghosh and Pathak (1962). The distribution/collection of different *Morchella* species in India is given in Table 1.1.

SYSTEMATIC TREATMENT

The genus *Morchella* is one of the oldest genera of Pezizales. It was established as early as 1719 by Dillenius. It has undergone little change till today regarding its limits. Though at times *Mitrophora* Lav. has been included in *Morchella* (Saccardo, 1889; Seaver, 1928 and Rifai, 1968), but majority of the researchers prefer to treat it as a separate genus, distinguishable by a pileus free from stipe in the lower region (Dennis, 1960, 1968 and Moser, 1963).

Opinions have differed with regard to the circumscription of the species in the genus *Morchella,* so that variable number of species have been recognized from time to time. Groves and Hoare (1953) and Dennis (1960, 1968) recognized three species, whereas Seaver (1928)

Table 1.1 Distribution of *Morchella* species in India

Species	*State*	*Reference(s)*
M. angusticeps **Peck.**	HP	Sohi *et al.* (1965); Munjal *et al.* (1977); Jandaik *et al.* (1978); Lakhanpal and Shad (1986)
	Kashmir	Ghosh and Pathak (1962); Batra and Batra (1963); Kaul (1976); Waraitch (1976)
	Maharashtra	Wakode (1963)
M. conica **Pers.**	HP	Sohi *et al.* (1965); Munjal *et al.* (1977); Lakhanpal and Shad (1986)
	J & K	Batra and Batra (1963); Waraitch (1976); Kaul (1976)
	UP	Hennings (1901)
	Maharashtra	Ghurde and Wakode (1981)
M. crassipes **(Vent.) Pers. ex Fr.**	HP	Lakhanpal and Shad (1986)
	Kashmir	Waraitch (1976)
M. deliciosa **Fr.**	HP	Munjal *et al.* (1977); Lakhanpal and Shad (1986)
	Kashmir	Ghosh and Pathak (1962); Batra and Batra (1963); Kaul (1976); Waraitch (1976)
	Maharashtra	Wakode (1983)
	UP	Ghosh and Pathak (1962)
M. esculenta **(L.) Pers. ex Fr.**	HP	Sohi *et al.* (1965); Lakhanpal and Shad (1986)
	Kashmir	Ghosh and Pathak (1962); Batra and Batra (1963); Waraitch (1976); Kaul (1976)
	Maharashtra	Wakode (1983)
	UP	Ghosh and Pathak (1962); Sydow and Butler (1911)
	Western Ghats	Kaviyasaran *et al.* (2006)
M. semilibera **DC. ex Fr. (Syn.)**	HP	Sohi *et al.* (1965); Lakhanpal and Shad (1986)
	Kashmir	Ghosh and Pathak (1962); Kaul (1976)

recognized six species and Boudier (1907) as many as thirty-two species. The delimitation depended upon the weightage one gives to the morphological variations. British mycologists follow a conservative concept and recognize six species, whereas American mycologists tend to split morels into a number of species and recognize more than six species.

The earliest record of *Morchella* from India is by Cooke (1870) as *M. gigasporsa* Cooke, but this species is now a synonym of *M. bohemica* Krombh (see Butler and Bisby, 1931) and presently it is placed in the genus *Verpa*.

All the six species considered as valid in the genus *Morchella* have been recorded from India. These are *M. deliciosa* Fr. (delicious morels), *M. esculenta* (L) Pers. ex Fr. (common morel), *M. conica* Pers. (conical morels), *M. crassipes* (thick stemmed morel), *M. angusticeps* Peck. (black morel) and *M. hybrida* (hybrid morel). The correct name for *M. hybrida* is *M. semilibera* DC. ex Fr. The former is untenable (Rifai, 1968) as it was not accepted by Fries (1822). The report of *M. vulgaris* Boud., which is a synonym of *M. esculenta* (L.) Pers. ex Fr.

from Assam by Bhattacharya and Baruah (1953) has been doubted by Ghosh and Pathak (1962) who have reported five above-mentioned species of *Morchella* from Kashmir. Sohi *et al.* (1965) reported *M. angusticeps, M. conica, M. esculenta* and *M. hybrida* from Himachal Pradesh. Tilak and Rao (1968) also listed *M. rotunda* which is presently treated as synonym of *M. esculenta* by Seaver (1942). Waraitch (1976) described an additional species, i.e. *M. crassipes* (Vent.) Pers. ex Fr. along with three previously recorded species from Kashmir, and as such the total number of valid *Morchella* species reported from India became six. Lakhanpal and Shad (1989) recorded all the six species from Himachal Himalaya along with a new species and a species *M. tibelica,* which had been reported only from China so far.

The six species concept is primarily based on mature fructifications, which have been keyed by Seaver (1942). The key for all the species recognized as valid is given below:

Key to Species

1. Base of the pileus free from stipe	*M. semilibera*
1. Base of the pileus adnate (attached) to the stipe	2
2. Ribs (ridges) of the pileus darker than the interior of the pits becoming smoky brown to almost black	*M. angusticeps*
2. Ribs of the pileus and interior of pits similar in colour	3
3. Pileus (ascocarp) small, 2-5 cm long and 1-2 cm broad	*M. deliciosa*
3. Pileus 7-10 cm long and 1-2 cm broad	4
4. Stipe conspicuously swollen at the base, pits large upto 4 × 1 cm, ribs thin (less than 1 mm in thickness)	*M. crassipes*
4. Stipe not conspicuously swollen at the base, pits not exceeding 2 × 1 cm, ribs thick	5
5. Pileus elongated and strongly attenuated upwards, apex acute	*M. conica*
5. Pileus ovate, apex obtuse	*M. esculenta*

NUTRITIVE VALUE

Mushrooms are low calorific but protein rich and hence a good source of protein for human consumption (Kurtzman, 1975). Although mostly regarded a delicacy, it would be more realistic to look upon them as a food which can contribute significantly in relieving protein deficiency. Normally, mushrooms contain 20-40 per cent protein on a dry weight basis and thus surpass many foods including milk in terms of protein content. Most species contain all the essential amino acids and most of them are found in about the same proportion as in egg. Mushrooms, like yeasts, are also rich in vitamins, with the exception of thiamine.

Species of *Morchella,* commonly called "morels", are still mainly collected from the wild. These have an excellent flavour, reputed to be superior to that of other mushrooms. Not only morels but even their mycelia have been reported to have the same nutritive value (Cochrane, 1958; Robinson and Davidson, 1959; Singer 1961 and Hayes and Hadded, 1976). Various workers (Kaul, 1978; Jandaik *et al.*, 1978 and Samajpati, 1978) have reported about chemical composition of morel fruiting bodies and/or morel mycelium. Mineral contents in pileus and stipe of *M. angusticeps* have also been recorded. Jandaik *et al.* (1978) have also given the proximate composition of morel mycelium produced in submerged culture. They reported the percentage of moisture, ash, fat, fibre, and protein content in two species of *Morchella* viz. *M. esculenta* and *M. deliciosa*. In both the species, the moisture content is 83.06 and 90.30%, while ash content is 0.96 and 1.12%, fat content is 2.97–3.13%, fibre content is 1.96–1.78% and the protein content is 34.70 and 29.16%, respectively.

The vitamin content (μg/g dry wt) of morel mycelium was also studied by Samajpati (1978) who recorded the presence of thiamine (4.16), riboflavin (17.6), niacin (80.6), pantothenic acid (9.4), pyridoxin (6.0) and B_{12} (0.004). The same author also stated that *M. esculenta* mycelium contained minerals as well. The concentration of calcium, magnesium and iron was found to be 3.16, 28.12 and 2.98 mg/100 g dry weight, respectively. Protein score for morel mycelium was recorded to be 40.1.

On the other hand, Kaul (1978) analyzed fruiting bodies of different *Morchella* species and reported following values in respect of various constituents viz. crude protein, fat, ash and moisture percentage. *M. angusticeps* has 31.06% crude protein, 3.0% fat, 16.82% ash and 90.13% moisture, while *M. conica* has 33.75% crude protein, 2.08% fat, 15.46% ash and 88.23% moisture. *M. deliciosa* has 34.37% crude protein, 2.75% fat, 14.88% ash and 89.17% moisture, and the last species analyzed, *M. esculenta* has 31.25% crude protein, 3.31% fat, 12.77% ash and 90.07% moisture, respectively. The contents of calcium, phosphorus, iron and potassium were also analyzed in the fruit bodies of various *Morchella* species by Kaul (1978) out of which *M. deliciosa* has maximum amount of calcium (0.7997 mg/100 g) and *M. hybrida* has the minimum (0.4371 mg/100 g). The phosphorus content is maximum in *M. esculenta* (3.313 mg/ 100 g) and minimum in *M. hybrida* (1.178 mg/100 g). Similarly, the iron content is maximum in *M. conica* (1.834 mg/100 g) and minimum in *M. hybrida* (1.056 mg/100 g). In case of potassium, the content is higher in *M. angusticeps* (3.954 mg/100 g) and lowest in *M. conica* (3.138 mg/100 g), respectively.

The pileus of *M. angusticeps* was also analyzed separately for contents of various macro- and microelements (Jandaik *et al.*, 1978). Pileus has greater concentrations of nitrogen (5.01%), phosphorus (1.20%), sulphur (0.39%), zinc (87 ppm), and copper (22 ppm), while stipe has greater concentration of potassium (3.0%), magnesium (0.12%), manganese (30 ppm) and iron (2.78 ppm), respectively. The biological value of a protein depends on the relative concentration of different amino acids in it, particularly essential amino acids.

There are many *in vitro* methods (chemical score, essential amino acid index and protein score) which are generally used to predict the nutritional quality and biological value of proteins. Samajpati (1978) has reported the presence of different amino acids in the mycelia of *M. esculenta* and *M. deliciosa*. The concentration of amino acids was found to range from 0.20 to 8.16 and 0.22 to 9.12 g/100 g protein in the mycelia of *M. esculenta* and *M. esculenta* and *M. deliciosa*, respectively (See Jandaik and Sharma, 1995).

Samajpati (1978) further stated that the contents of essential amino acids and sulphur containing amino acids ranged between 0.2 (cystine) and 5.16 (leucine) g/100 g protein in mycelium of *M. esculenta*. Supplementation of human diets with morel mycelium would go a long way in overcoming the protein malnutrition problem in developing countries like India.

Shad (1989) analysed six species of morels for moisture, protein content, carbohydrates, reducing sugars, non-reducing sugars, starch, non-essential free amino acids and amino acids. The moisture content of the six species of *Morchella* was observed to be 84.15–88.14%

on fresh weight basis. The total soluble carbohydrates, reducing sugars, non-reducing sugars and starch were found to be 21.0–27.2, 1.3–2.8, 1.95–25.5, and 21.1-18.3 per cent on dry weight basis in carpophores of all six species. The morel sporophores were observed to be rich in phosphorus, potassium, and magnesium and poor in iron. The non-essential free amino acids were found to be 6.6–11.8 per cent in the sporophores of the six species. The amino acids analyzed from the six species are arginine, alanine, aspartic acid, cystine, glutamic acid, glycine, isoleucine, leucine, methionine, phenylalanine, proline, serine, threonine, tyrosine, valine and tryptophan.

Sharma (1993) analysed the nutritive components of seven species of *Morchella* viz. *M. angusticeps, M. conica, M. esculenta, M. deliciosa, M. crassipes, M. tibelica,* and *M. rotunda*. The total carbohydrate content was observed to be highest in *M. crassipes* (61.65%) followed by *M. deliciosa* and others. It was lowest in *M. tibelica* (45.8%). The fat content in morel species ranged between 1.1–2.9%; *M. deliciosa* (2.9%) possessed the highest amount and *M. angusticeps* the lowest (1.1%). All the species were observed to be rich in fibre content. In most of the morel species it ranged between 14.5–16.6%. Crude protein content varied between 11.38–21.9%. The ash content varied from 8.7 to 9.11%, among different morels. The highest ash content was observed in *M. esculenta* (9.11%). The energy value on dry weight basis among morels ranged between 29.0 kcal/100 g and 321.80 kcal/100 g with *M. crassipes* (321.80 kcal) having the high energy value and *M. tibelica*, the lowest.

The potassium, phosphorus, sodium, and calcium content varied in different morel species from 1770–2400 mg, 210–330 mg, 80–190 mg, 135–160 mg/100 g on dry weight basis, respectively. As for some microelements, the content ranged from 50.7 to 105.7 mg/100 g (iron), 79.6–11.6 mg/100 g (magnesium), 10.2–1.8 mg/100 g (Zinc), whereas manganese and copper were present only in negligible amounts. Nine amino acids were identified among the members of morels viz. arginine, histidine, alanine, cystine, lysine, tyrosine, valine, isoleucine and leucine. Alanine, lysine, arginine and histidine were observed to be the dominant amino acids in free amino acid pool.

ECOLOGICAL ASPECTS

Among the physical factors in the natural environment affecting the larger fungi, the important ones are moisture, composition of substrate, pH, available nitrogen and mineral status (Grainger, 1946). Munjal *et al.* (1977) and Kaul *et al.* (1981) also emphasized the significance of some of these factors in the appearance of morels in Himachal Pradesh and Kashmir valley, respectively.

Morchella species occur in association with species of vascular plant so that a survey of both types of species is necessary. Various localities within Himachal Himalayas have been explored for morels by Shad (1989).

With regard to morels, Mc Cubbin (1913); Moser (1949a,b); Ramsbottom (1953) and Baker and Matkin (1959) reported excessive growth of *Morchella* on disturbed ground, burnt places and where burnt ash of paper or wood had been added. Baker and Cook (1974) investigated that fruitings of morels in spring may be due to their ability to grow at low temperatures to the exclusion of competition. Delmas and Bunel (1974) pointed out (after Heim, 1936 and Cailleux, 1956, 1969) that one of the difficulties in cultivating morels lies in the ecological specificity of each species of morels. They provided a list of basic factors on ecology of morels, e.g. types of soils, associated trees, altitude, season of appearance, etc. and emphasized that their trials on various reconstituted natural media had yielded no fruiting bodies (see Chang and Hayes, 1978).

Schmidt (1983) studied the life cycle and ecology of morels in the field. He observed that the ascospores of *Morchella esculenta* germinated at temperature as low as 2°C *in vitro*, and the spores buried in morel sites in cellophane bag swelled but did not produce germ tubes until the soil temperature exceeded 10°C, approximately. But the spore germination was inhibited as the soil continues to warm up after fruiting body emergence above 10°C and the spores did not remain viable after one year near the soil surface. He explained that sterile rye buried two weeks in the field was readily colonized by *M. esculenta* when cased with site soil containing either spores or sclerotia when soil temperatures remained below 10°C, but as the temperature increased, only other saprophytes were isolated. He also observed that heterokaryon formation could occur at this stage in the life cycle of *M. esculenta*.

Kaul (1975) analyzed forty soil samples from morel bearing and burnt situations for major inorganic components, organic matter and moisture to determine the factors responsible for carpophore formation. Correlation of the results obtained indicated that higher calcium contents of soil and alkaline reaction of the substrate may have a role in carpophore development. Munjal *et al.* (1977) studied the edaphic factor in relation to growth of *Morchella* species. They stated that *M. deliciosa* is adapted to low elevation zone and prefers loamy soil, while *M. angusticeps* and *M. conica* grew at higher elevations, indicating that they require cooler climatic conditions and more aeration. The analysis of the soil samples of different substrates shows that *Morchella* species prefer higher pH (8.1); this supports the view of Repin (1901) who claimed to get more fruiting bodies of *Morchella* around a bed composed of dry leaves of apple deposited in a trench and rendered alkaline with sodium carbonate.

Kaul *et al.* (1981) investigated the morel bearing sites in Kashmir valley and noted that morel fruit bodies were found when soil moisture was above 20%, soil temperature ranged between 10-23°C above ground and 8.5-20°C below ground, relative humidity 58-77%, pH neutral or alkaline and when there was high level of available nitrogen. They also observed higher levels of calcium, potassium and phosphorus in the soil during the fruiting season.

Lakhanpal and Shad (1986) and Shad (1989) attempted to study and coordinate the different ecological aspects of morel bearing areas. They made observation on seasonal distribution of morels and their association with different plants. They studied the relation of *Morchella* production to forest type, the fruiting period of morels plus parameters such as temperature, pH and NPK and Ca contents of morel bearing and non-morel bearing soils. The maximum production of *Morchella* has been observed to be in the temperate mixed coniferous forests, especially in deodar forests, from March to April, depending upon the amount of snowfall and subsequent precipitation. The colour of the species has been seen to vary with soil and forest types. *M. esculenta* has been collected at all altitudes from 1600 to 2800 m, *M. crassipes* and *M. conica* at middle elevations, *M. deliciosa* and *M. angusticeps* at lower and higher altitudes, respectively. Therefore, minor colour variations should be used with restraint while delimiting the species.

Temperature at surface level at 0-12 cm of the collecting sites ranged from 2–15°C and from 10–20°C, respectively, during morel season. Moisture and relative humidity of morel bearing soils are ranged from 18.29–33.09% and from 54–94% respectively, their pH has been found to be slightly alkaline and NPK and Ca content of morel bearing soils have been found to be high as compared to non-morel bearing soils.

RHIZOMORPHOSPHERE ASSOCIATIONS

Since rhizosphere microorganisms greatly influence the root activity, pioneering efforts were made to know whether similar activity and effect is exercised by microorganisms on the rhizomorphs of *Morchella* species or not. To substantiate this hypothesis, the mycoflora associated with rhizomorph of *Morchella* species was studied and the zone of microbial rhizomorph was termed 'Rhizomorphosphere' (Shad, 1989 and Lakhanpal and Shad, 1991).

From the rhizomorphosphere of morel bearing soil, forty-eight species of soil fungi and some forms of sterile basidiomycetous mycelia were isolated. Out of these, twenty belonged to Deuteromycotina, sixteen to Ascomycotina, ten to Zygomycotina, and some were white sterile mycelial forms belonging to Basidiomycotina.

Trichoderma viride, T. harzianum, Penicillium chrysogenum, Rhizopus nigricans, Alternaria alternata, Aspergillus flavus, A. fumigatus, A. luchuensis, Cephalosporium roseum-griseum, Verticillium sp. and white sterile mycelium were the most common species in the forest soils bearing six *Morchella* species. Out of these, *T. viride* was associated with all *Morchella* species and in all forest types. Indices of similarity for the four forest types and for the six species of *Morchella* have also been worked out following Sorensen (1948). The indices of similarity were highest for the tropical and subtropical forest types and least for the tropical and mixed coniferous forest types in all the species. The index of similarity was highest for *M. esculenta* and *M. crassipes* followed by *M. crassipes* and *M. deliciosa, M. deliciosa* and *M. esculenta* and *M. esculenta* and *M. conica*. The index of similarity was lowest for *M. semilibera* and *M. crassipes*. These low values presumably show that *M. semilibera* is distinctly different from the other species of *Morchella.*

ETHNOMYCOLOGICAL SIGNIFICANCE

Among the large number of fleshy fungi that inhibit North-West Himalayan ranges in Himachal Pradesh, *Morchella*, the true morel, is one of the most important fungus from economic, social and ethnomycological point of view. Lakhanpal and Shad (1986 a, b) conducted ethnomycological studies on morels of N-W Himalayas. In these studies people were interviewed in the morel producing areas to gather information about collection of morels, myths associated, propagation in nature, local and vernacular names of morels, and recipes of morels used by local people.

It was revealed that morels are collected by a variety of people, shepherds, herdsmen, rural women and girls, school-going children and the rural poor peasants who spend most of the time in morel collections during the morel season. Morels, known as 'Guchhi' are known by different local names, e.g. 'Chiaun' in Shimla district which becomes 'Kiaun' in Kullu valley. In some areas in Kullu and Mandi, they are also called 'Dunglu' and 'Chunchuru' in Kinnaur and 'Chamkaid' and 'Kankenadeu' in Chamba districts.

Many myths are associated with morels among the local inhabitants. The most common belief is that morels emerge from ground after lightning strikes it and that they come up in greater abundance in the burnt up areas. Also that only few lucky or naturally gifted people can collect them. There is a common myth among people that *Morchella* sporophore does not increase in size, once it is spotted.

This has led to indiscrete collections, removing even the smallest bits of fructifications. Some wise men, however, crush one or two fruiting bodies at the collecting sites ensuring inoculum for the following seasons. Such types of practices need to be followed to maintain continuity of inoculum in nature. Deforestation has also led to the loss of inoculum at different sites.

Morchella has many recipes and it is consumed differently in different localities, but the commonest method of its consumption is by preparing Pulao of morels with rice, locally known as 'Chiaun ka Bhat'.

PRODUCTION AND TRADE

An estimated 60 ± 5 tonnes of dry morels are exported every year from India to the international markets. Himachal Pradesh, Jammu and Kashmir and Uttar Pradesh are the main exporters of morels in India. The morel production during 1981-82 from Himachal Pradesh was around 11000 kg, whereas during 1982-84 it was stated to be 48000 kg (information collected from Forest Department). The production figures for 1986-87 have been estimated to be 55000 kg. Sociobiological studies conducted in morel producing areas of Himachal Pradesh by Shad and Lakhanpal (1991) revealed that around 80,000 kg of morels are collected/produced from Himachal Pradesh. On an average, it has been estimated that morels add Rs. 500 to 5000 to the family income of morels collecting sectors. The forests are presently leased out to individual contractors by the Forest Department for morel collection, and morel collection by others is prohibited. However, in practice, especially in Shimla district, many sections of the society, viz. village women group, school-going children, Nepali labour and many others also collect morels, which are offered for sale either in fresh or dry form to the local shopkeeper or middle men. The price varies from place to place and, in general, Rs. 700–1000/kg dry produce is offered to villagers. The middle men sell their produce to the wholesaler at the rate of Rs. 1200 to 1500 per kg and finally it reaches the consumer at the rate of Rs. 3000—4000/kg dry. At present, the market retail price for dry morels in Solan and Shimla markets ranges between Rs. 5000 and Rs. 8000 per kg dry (Jandaik and Sharma, 1995), and after some years it may be even Rs. 10,000 per kg (personal observations).

Obviously, the consumer has to pay three to four times more price than the price given to the collector. The huge benefit is taken by the middle men and this can easily be avoided provided that morel market is channelized by opening morel cooperatives, nationalization of morel collection and declaring support price to morel crop coupled with proper licensing policy.

Detailed observations on these and other aspects of morel biology are presented in the following chapters beginning with life cycle pattern, systematic treatment, molecular characterization, physiology, ecology, association with soil microflora and higher plants, ethnomycology, sociobiology, cytology of the ascus development, nutraceutical and nutritional potential, various consumption patterns and traditional and modern recipes.

CHAPTER

2 The Life Cycle Pattern

CHAPTER OUTLINE

- The Ascocarp
- Ascospores and Mycelium
- Heterokaryon Formation
- Sclerotial Development
- Ascocarp Development
- Cultivation Aspects

Different aspects of morel biology have been well worked out all over the world. Orientation of every study has been towards development of technology for artificial cultivation of morels. But foolproof method for commercial cultivation of morels still eludes scientists. The life cycle, however, is well studied and various steps in the life cycle seem to be as follows: The ascocarp, the morel of commerce, is the fructification that appears in nature and is the article of commerce. It has sterile ridges and fertile grooves or depressions. The latter bears hymenium which is lined by asci containing ascospores. Ascospores on germination produce mycelium, which becomes heterokaryotic, forms sclerotia and then produces ascocarps. The 'unseen' thread that binds the different events in the life cycle in morels is still not completely traced, though lot of work has been done on individual 'events', which is being reviewed here. Once the 'missing link' is discovered, the morels, no doubt, will be 'booming' and 'mushrooming' in the homes.

THE ASCOCARP

The fructification or ascocarp in *Morchella* is stipitate, hollow and differentiated into pileus (cap) and stipe (stalk). The colour of the pileus varies from dirty greyish-white to dark-brownish depending upon the species, age of fruiting body, and the type of vegetation. Pileus constitutes the fertile portion of the ascocarp. It is hollow and fleshy. The honeycombed surface of pileus gives the appearance of a sponge. When young, the surface of pileus is quite smooth, and as it grows, there appears a network of ridges and grooves or depressions (pits) on the surface due to unequal growth of the hymenial surface. The ridges are sterile while the pits are fertile.

Pileus is variously shaped in *Morchella* species. In *M. angusticeps,* pileus is elongated, sub-globose, or elongated with slightly subconic apex and adnate at the base; in *M. conica,* pileus is elongated, more or less conical in shape, acute at the apex, subobtuse, also adnate at the base; in *M. crassipes,* pileus is more or less conical in shape and adnate at the base; in *M. deliciosa,* pileus is elongated to sub-globose, obtuse at the apex, adnate to the stipe base; *M. esculenta,* has sub-globose ovate or elongated pileus attenuated upwards but obtuse at the apex and adnate to the stipe at the base; in *M. semilibera,* the pileus is conic to elongated, free at the base with acute or obtuse apex light brown in colour; in *M. tibelica* also the pileus is elongated and conical with conic apex and adnate base stipe; and in *Morchella simlensis,* the pileus is globose or elongated with obtuse apex and adnate base stipe.

ASCOSPORES AND MYCELIUM

The hymenium which lines the pits bears numerous asci. These asci are cylindrical, sub-cylindrical with obtuse apex. Each ascus contains eight ascospores. The ascospores are ellipsoid, uniseriate, yellowish in mass, hyaline singly, smooth and eguttulate but with external oil drops at each end. The asci are positively phototrophic. The ascospores are wind disseminated. Falling on a suitable substratum (moist humus), each ascospore germinates to produce new mycelium. The ascospore that happens to fall on unsuitable substratum will perish (Shad, 1989). Mycelium is inconspicuous and subterranean. It grows few inches deep in soil. The mycelial colonies are white, turning white-brown with age.

Volk and Leonard (1990) studied the germination and cultural characteristics of hyphae in morels. They raised monosporous cultures and observed development of copious vegetative mycelium which fills up a Petri plate of 8.5 cm at 22-25°C in just 5-6 days. The mycelium secretes a dark brown pigment into the medium and the browning starts from the centre and proceeds to the periphery with age and mature colonies turn brown in colour. The morel hyphae have relatively large diameter (5-10 μm), branch and anastomose frequently and are multinucleate with 10-15 to 40-50 or even upto 65 nuclei per cell. This type of multinucleate condition in morels is exceptional among ascomycetes (Volk and Leonard, 1990). These characteristics of hyphae may be required by morels because of the large fruiting body which involves substantial channelling of nutrient reserves into the developing ascocarps. It may, however, be pointed out that natural mycelium forms differ morphologically from the cultural ones and normal mycelium is not able to survive frost (Buscot, 1993).

Hervey *et al.* (1978) observed 100% germination of ascospores on 2% Malt Extract Agar and 0.3% Yeast Extract Filter Paper Agar at 20°C after incubation for 24 hours. Schmidt (1983) carried out studies on spore germination of *M. esculenta in vitro* and in natural soil and reported that germination takes place at temperatures as low as 2°C. Under natural conditions germination takes place in the range of 10-15°C. Ascospores in soil germinate during limited period that may be indexed to the temperatures generally found during the formation and maturation of ascocarps. Complex nutritive, physical and biotic changes that occur in soil are also considered significant for germination. Schmidt (1983) concluded that *M. esculenta* is a psychrotolerant fungus, which reinforces the earlier observation of some authors that the fruiting of morels in spring may be due to their ability to grow at low temperatures to the exclusion of competition. At the end of sporulation period, which lasts a few days in morels, the ascocarps undergo a rapid autolysis and disappear (Buscot, 1989)*.

HETEROKARYON FORMATION

The failure to cultivate morels is ascribed to the use of monoascosporus cultures. Hence, the role of heterokaryosis in the life cycle of morels needs to be thoroughly investigated. Hervey *et al.* (1978), Volk and Leonard, (1989 b, 1990) suggest a significant role of heterokaryosis in the life cycle and formation of fruiting body of morels in the laboratory. The occurrence of heterokaryons is widespread in higher groups of fungi, e.g. ascomycetes, basidiomycetes and fungi imperfecti. Heterokaryosis is known to confer the potential for continuous variation in fungi. Heterokaryons arise either by mutation of genes within one of the existing nuclei or by hyphal fusions between genetically different colonies. An important consequence of heterokaryosis is that recombination can occur even in homothallic fungi. Ower *et al.* (1986) pointed out that morels may be homothallic and fruiting may involve either autogamous or heterogamous nuclear pairing prior to meiosis in the fruiting bodies. Schmidt (1983), while observing germination of ascospores in *Morchella esculenta,* observed numerous instances of

*Our observations are exactly contrary to those of Buscot (1989). The sporocarps have been very rarely observed to decompose in nature through autolysis except when these fall to the ground. Otherwise these shrink, dry but remain intact.

direct hyphal growth from a germinating spore to the ungerminated spore and suggested that heterokaryon formation could occur at this stage in the life cycle.

Hervey *et al.* (1978) crossed several hundred pairs of single ascospore isolates of *Morchella esculenta.* They observed the build up of aerial hyphae at the interaction interface in all non-self combinations and termed this as "barrage-like build up". They observed that either ascus fusion nuclei must be heterozygous for certain genetic factors or that asci of a single ascocarp do not at all arise from a single pair of nuclei. They further hypothesized that heterokaryosis must occur at some time in the life cycle because of the observed segregation patterns of two morphological cultural characteristics (flat and fluffy) among progeny of a single fruiting body.

Volk and Leonard (1989 b) taking clue from reports of Fincham *et al.* (1979) on *Neurospora* and *Podospora* that heterokaryosis is under the control of multiple genes, including in some cases, the mating type genes, wanted to further characterize the barrage type build up of aerial hyphae cytologically, culturally and genetically. They showed that *Morchella* is capable of forming a heterokaryon between monoascosporus strains. In cultural evidence, they noticed that at the point of mating of hyphae raised from spores of the same species in the same Petri dish, an aerial bridge of hyphae [referred to as "barrage-like build up of hyphae" by Hervey *et al.* (1978) and retermed as "mycelial meld" by Volk and Leonard (1989)] was formed when two colonies met in majority of the cases followed by deposition of a dark brown pigment into the medium. The cultural characteristics and mycelial interactions between two genetically dissimilar cultures suggested presence of heterokaryotic phase in the life cycle of morels. But whether heterokaryotic phase is essential in the morel life cycle was not demonstrated. Further, no mycelial meld was noticed for reaction between the mycelia from the same spore or between genetically identical sister spores. Two different mutations isolated in this study were used to show genetic complementations in the heterokaryons. Interspecific hyphal fusion has also been reported among *M. esculenta, M. crassipes,* and *M. deliciosa* but not with *M. semilibera* and *M. angusticeps.*

Cytological studies of aerial mycelial meld demonstrated nuclear pairing in presumptive heterokaryon. Volk and Leonard (1989b) suggested that mycelial meld is a point at which heterokaryosis takes place in the morel, and evidence obtained indicates that a stable heterokaryosis takes place in the morel and a stable heterokaryon is formed in the melds. But whether or not this heterokaryon with its paired nuclei is associated with sexual reproduction or is strictly a vegetative trait remains to be elucidated. Ability to process mutants from the meld reaction through sclerotia and to the fruiting body would reconcile this problem (Volk and Leonard 1989b).

Volk and Leonard (1990) maintain that there is no firm evidence that nuclear pairing is a natural phase of the morel life cycle or even that two pairing nuclei are in fact different. However, this nuclear pairing is a common feature of forced heterokaryotic mycelia in *Morchella* and is also seen in certain sclerotia and in the sterile cells of the fruiting body, providing a possible link between heterokaryon, the heterokaryotic sclerotium and fruiting body formation. Buscot (1993) also observed no evidence of mating types in morels.

Volk and Leonard (1989) postulated that generations of forced heterokaryons under laboratory conditions and their processing through sclerotia to fruiting body should lead to reproducible and more dependable method for morel cultivation. Sehgal and Sharma (2007) studied forced heterokaryons formation in *M. esculenta* and observed pairing of nuclei in the mycelia derived from ascocarp and generated heterokaryons, indicating a possible role of heterokaryosis in the life cycle of *Morchella*.

SCLEROTIAL DEVELOPMENT

The significance of sclerotia in the life cycle of morels was first realized and emphasized by Ower (1982) and Ower *et al.* (1986). They demonstrated that sclerotia have a central role in the production of fruiting bodies under controlled conditions. They showed that *M. esculenta* sclerotia act as nutrient sink from which fructification during development drains nutrition. Since then, number of workers have made significant contribution in the need for determining optimal conditions for the formation, growth and maturation of sclerotia and for the regulation of their production (Buscot and Roux, 1987; Buscot, 1989, 1992, 1993; Volk and Leonard, 1989 a, b, 1990; Buscot and Bernillon, 1991; Buscot and Kottke, 1990 and Amir *et al.*, 1992, 1993, 1994, 1995 a, b).

In nature, sclerotia exactly similar to those formed in cultures, have not been observed so far. There are formed some subterranean mycelial structures akin to those in cultures. Sclerotia obtained in culture are not true sclerotia in the classical sense with tissue differentiation, but are more akin to the undifferentiated pseudosclerotia of some Ascomycotina.

The formation of sclerotia was first reported in morels almost simultaneously by Molliard (1905) and Fron (1905). The former observed their formation on 'sterile moistened bread' and considered them to be imperfectly developed fruiting bodies, whereas the latter reported them to be large, looking and tasting like fruiting bodies. Both of them, however, did not realize the significance of these structures in the life cycle of the morels.

Development of sclerotia in morels has also been reported by Robbins and Hervey (1959) and Kaul *et al.* (1981). The former recorded that *M. crassipes* formed sclerotia on 2% Malt Agar, whereas the latter found significant difference in the formation of sclerotia in eight different strains of *Morchella* representing six different species and using two synthetic and two semi-synthetic media. None of the strains were found to produce sclerotia on two synthetic media used in these studies. Some of the cultures (*M. esculenta* II and *M. deliciosa* II) failed to develop sclerotia even on any semi-synthetic media (PDA & MA). Mehta and Sharma (1992) also observed sclerotia formation by different species of morels on PDA and YPDA.

Volk and Leonard (1989a) in their studies on sclerotia formation in *Morchella crassipes* used a modification of Jar method of Ower *et al.* (U.S. Patent–1986). They found that rye grain supplemented with peptone, yeast extract, trace elements and Casamino acids overlaid with perforated aluminium foil and covered with a layer of nutrient poor soil (soil consisted of 3 parts potting soil mixture, 1 part peat soil and two parts distilled water with an initial pH of 5.8) gave the best results. Addition of asparagine or aspartic acid as a nitrogen source to the rye also had a beneficial effect on sclerotium formation, while addition of carbon sources had no such effect. Maximum sclerotia were produced in continuous darkness; continuous light, however, completely inhibits sclerotium formation. However, the sclerotia were small.

In the subsequent studies on *Morchella esculenta*, Volk and Leonard (1990) observed that sclerotia formation in mycelial culture occurs when it is grown at low temperature (4°C) or on a complete medium (glucose peptone yeast extract, mineral salt medium plus 2% compost sheep manure) or when nutrient depletion occurs (such as complete colonization of a Petri dish).

The development biology of sclerotia formation in different groups of fungi has been reviewed by Wiletts and Bullock (1992), but unfortunately morel sclerotia are not included in their review. Studies on nutritional requirements of sclerotia formation have been done only by a few workers.

Small 1-2 mm sclerotial initials begin to form after 7-10 days growth which subsequently expand to form a single large sclerotium. Sclerotia are considered mature when radial expansion ceases and the pigmentation attains a dark brown colour. Sclerotia are the terminal type formed from the repeated branching of a terminal hypha. However, Amir *et al.* (1992) maintain that *Morchella esculenta* sclerotia initials appear laterally to the main leading hypha and thus can be classed as lateral type.

Volk and Leonard (1990) further observed that during the formation of sclerotia, the cells round up to form varied and unusual shapes with thick walls and remain multinucleate, store nutrients and accumulate oil droplets. A cross-section of sclerotia reveals a series of compacted, isodiametric cells with very thick wall, able to tolerate adverse conditions such as low temperature and desiccation.

Intensive studies on morphology and physiology of *Morchella esculenta* during sclerotial formation were carried out by Amir *et al.* (1992, 1993, 1994, and 1995 a, b). One of the critical requirements for culturing *Morchella* spp. in a two-layer system is the production of large sclerotial biomass in the poor substrate. In order to understand the process and factors involved, Amir *et al.* (1992) devised an improved experimental system—split-plate procedure. In this method, Petri dishes (9 cm in diameter) were divided into two by a plastic barrier to prevent diffusion. One side of this split-plate contained nutrient poor medium (Noble Agar GNA), while the other side contained nutrient rich medium (PDA, DM, etc.). Inoculum placed on the poor side grew towards the rich side, ultimately producing sclerotia mainly on the poor medium. They also studied the effect of addition of salts, sugar and PEG and resultant water potentials in the poor medium. While salts inhibited the production of sclerotia, sugars especially hexoses and hexitols enhanced it. They concluded from their experiments that two conditions are required for the production of the medium—a turgor potential in the hyphae conducive to morphogenesis and the type of sugar added to the poor medium. Maximum sclerotium formation was obtained on the noble agar side (GNA) of the plate when 0.5 M hexose was added.

Amir *et al.* (1993) identified six major stages in the growth of sclerotia in *M. esculenta*. The first two stages refer to the colonization of the plate by the hyphae and the other four stages to the formation and development of sclerotia on GNA side. The most prominent changes on the GNA side occurred when the hyphae on the DM side reached the end of the plate. Restriction of the linear growth stimulated the development of new branches from tertiary branches.

Restriction of the linear growth also reserved the cytoplasmic stream when it started flowing from the DM side to the GNA side—point of origin of sclerotial initials. This reverse translocation (from younger to the older parts of colony) is not found in any other sclerotium forming fungus and is the characteristic of *Morchella* only. Another discovery made by them was that the anatomy of leading hyphae adjusts to rapid stream of nutrients from the mycelia to the sclerotia by development of long shallow pipes on the GNA sides. The results of their study also showed that sclerotia formed on the GNA side serve as a nutrient sink whose source is the mycelium on the DM side. This source is exhausted by sink as the sclerotia mature.

Amir *et al.* (1994, 1995 a) made measurements of glucose translocation during growth and sclerotial formation and observed the existence of source-sink relationship between mycelium and sclerotia. These studies suggested that the mechanism of translocation consists of turgor driven mass flow. Amir *et al.* (1955 b) also proposed a model for translocation during sclerotial development and showed turgor pressure gradients in fungi between source and sink by direct measurement. These studies also indicated the presence of turgor gradient inside the sclerotial tissue (Kaul, 1997).

Mehta and Sharma (1992) reported that among the ten natural and semi-synthetic media, sclerotia were formed only in Malt Extract, Glucose Asparagine, Perti's mineral, Richards and Leonian Agar medium. In liquid solution under *in vitro* conditions, no correlation could be worked out between mycelial growth and sclerotial formation, and the formation of sclerotia was specific to the basal media used in all species.

Further, significant finding in the sclerotial morphology was made by Buscot (1993) while studying mycelial differentiation of *Morchella esculenta* in pure culture. Observing the development of mycelia derived from single or several ascospores of *M. esculenta* on two semi-synthetic media at different temperatures, they found the development of two kinds of sclerotia. The distinction between the two was clear on polyspore culture on medium A (agar-10 g; malt extract-5 g and glucose-10 g/l) at 25°C, and they were termed as 'early encrusting sclerotia – EES' and 'late isolated sclerotia- LIS'. The first kind of sclerotia formed abundantly and synchronously at the colony margin 2 days after reaching the edge of the Petri dishes (respectively 6 and 8 days on 9 cm and 14 cm dishes). The sclerotia did not exceed 0.2–0.5 mm diameter individually but tended to aggregate into circular crusts, which rapidly become pigmented. The second kind of sclerotia (LIS) began to form, irrespective of dish size upto the 12th day after inoculation. They remain separated, could occur anywhere on the mycelium and were few in number (7-8 per mycelium). Their central part became pigmented while their margins remained growing, attaining 5-8 mm diameter. Out of 95 single ascospore cultures on medium, none produced EES and only 8 produced several LIS, which were initiated after 6 weeks.

The pattern of sclerotium production on medium B (agar-10 g; yeast extract-4 g; starch-15 g; glucose-10 g; KH_2PO_4-1 g; $MgSO_4.7H_2O$-0.05 g and pH adjusted to 6.5) differed from that on medium A. Most polyspore cultures developed radial lines of aerial mycelium delimiting from 2 to 6 irregular sectors, on which both, only on or even none of the two types of sclerotia formed. Sectoring occurred only in polyspore cultures and not on monosporal ones. EES initiation is promoted by growth interruption, whereas LIS initiation is related to the ageing of the culture. Considerable variation in the development of aerial secondary hyphae was noticed on the two media with single ascospore or polyspore culture. In all cultures, formation of abundant aerial secondary hyphae, occurring only with sclerotium production, was reduced or delayed suggesting that these processes are competitive and distinction between 'flat' and 'fluffy' may not be absolute. The earlier observation (vide Buscot 1993) that morel sclerotia form only after the total available nitrogen has been depleted from the medium would suggest that limitation on supply of this nutrient may fundamentally affect the balance between aerial mycelium formation and sclerotium formation. The ability to shift from one pattern of development to another depending on nutrient availability may be significant during natural colonization process (Kaul, 1997).

ASCOCARP DEVELOPMENT

Fructification of mushroom is believed to progress through four stages: initiation leading to the formation of primordia; maturation during which primordia develop into mature sporocarps; sporulation and autolysis or disintegration (Manachere, 1974). Buscot (1989) carried out field observations on growth and development of *Morchella rotunda* and *Mitrophora semilibera* in relation to forest soil temperature. He found that maturation in both lasted for 3-4 weeks. During this time ascocarp development progressed in three different sub-stages:

(i) Pre-emergence—ascocarps are still covered by litter. They are small (2-15 mm) and develop a distinct stipe and a furrowed hymenium. Pre-emergence may last for 2 or 3 weeks during which the growth rate is very slow (1-2 mm every 2 days).

(ii) Emergence—there is sudden increase in ascocarp growth rate. The ascocarps rapidly emerge from the litter and reach two-thirds of their final size within 24 hours which gives the impression that fruiting bodies develop overnight.

(iii) Final maturation—after rapid expansion, the rate of ascocarp development decreases and finally fruiting bodies reach their mature size and pigmentation and begin to sporulate.

Buscot (1989) further obtained data, which correlates the initial extent of spring time reheating of soil with ascocarp maturation. Volk and Leonard (1990) obtained the development of fruiting body from tiny primordium to mature size in incidental culture with begonias (*Begonia tuberhybrida*). The primordium is at first a tiny white fungus-like mass of hyphae. The hymenium differentiates at the very apex, enlarging and pigmenting as the primordium grows into a mature fruiting body (c.f. Kaul, 1997).

CULTIVATION ASPECTS

Mycologists want to cultivate morels just like shiitake or button mushroom and there have been numerous attempts to achieve success ever since morels have been consumed. There are various reports in literature which mention the attempts to cultivate morels. Initially, attempts have been made to reproduce natural conditions under which the fungus grows. Ramsbottom (1953) cites that the probable earliest mention to grow morels artificially was perhaps by J.G. Gleditsch in 1753. He mentions that countrywomen of Neo Marchia noticed appearance of morels at burnt sites. This observation led to burning of areas for morel production. But this practice had to be abandoned as it resulted in frequent forest fires. That morels appear on burnt grounds abundantly has been reported subsequently by a number of workers (McCubbin, 1913; Krieger, 1967; Moser 1949 a, b and Backer and Matkin, 1959).

Though Costantin and Matruchot were the first to start experiments on the cultivation of *Morchella esculenta* in France in 1883 (vide Costantin, 1936), Costantin repeated Milliard experiments with *M. hortensis* and obtained a yield of 350 g of fruiting bodies/sqm of bed surface. Yet Baron d'Yvoire (1989) was the first to succeed in growing these fungi. He obtained fruiting bodies of *M. esculenta* on apple rinds + humus + pieces of ripe morel carpophores taking clue from Roze (1883) that *M. esculenta* parasitizes the rhizomes of Jeurasalem artichoke—*Helianthus tuberosus*. This method has been outlined by Baker and Matkin (1959) as follows:

"Starting from a plantation of artichoke in May or June, one incorporates into the soil pieces of morels and waters it 4-5 times during the summer with dilute potassium nitrate. In autumn, apple pulp was spread around the artichoke stalks. One to two weeks later the bed of apple pulp was covered with a layer of dry leaves protected with twigs of oak or beech. In April or May the twigs and leaves were carefully removed and the area shaded with a cloth. A yield of 300 morels was obtained in the area of about 100 square feet. Fruiting occurred year after year from them".

Several authors have suggested other techniques and substrates. Repin (1901) got fruiting bodies around a bed composed of dry leaves rendered alkaline with sodium carbonate in a trench in which apple residue had been deposited. He also established that morels can grow saprophytically. Molliard (1904 and 1905) conducted trials in open ground under cherry tree and under greenhouse conditions on apple puree and discarded old papers by Molliard. He also experimented with *M. hortensis.* He obtained the fruiting bodies of *Morchella esculenta* on a bed of alternate layers of apple residue and paper waste in open ground under cherry trees covered with knee deep earth. Subsequently, he repeated the success experiments with *M. hortensis*.

Matruchot (1909 and 1910) also obtained *M. esculenta* fruiting bodies on paper pulp waste and rotten wood, mixed with sand. Heim (1936) successfully obtained the fruiting bodies of *Mitrophora semilibera* (a species closely related to morels) on substrate made of leaves and moistened soil. Heim (1936) claimed that there is no problem in cultivating morels and it has been demonstrated successfully that morels can be grown artificially, especially *M. hortensis*. However, what is missing at present is a contribution of experiments leading to better understanding of the biology of *Morchella* species that would help in developing a process for morel cultivation which can easily be reproduced by other workers as well.

Furthermore, it has been observed that morel mycelium forms an imperfect fructification, which has been isolated as fungus imperfectus and called *Costantinella cristata* (Matruchot, 1909 and 1910). A similar imperfect form undoubtedly also occurs in *M. hortensis* (Singer and Harris, 1987). Whether other species of the genus *Morchella* also produce the same imperfect state (as observed in case of *M. esculenta* and *M. hortensis*), warrants detailed investigations.

The use of pulverized rooted pine wood or sawdust of poplar and leaves, twigs, roots and wood of deciduous trees have also been used in cultivation experiments (Boyer 1891; Heim, 1936 and Falck, 1920). These techniques, however, remained fairly odd and hazardous due to lack of knowledge on the biological cycle and ecology of morels.

Success in controlled cultivation of morel fruiting bodies has been achieved in the recent past. A new direction to the experiments on morel cultivation was given with the publication of a note on the development of morel ascocarps by Ower (1982). Working in mycology laboratory at San Francisco State University (USA), he obtained 16 mature ascocarps from 8 cultivations. Experiments were conducted in walk-in growth chamber maintained at 15-18°C and 85% relative humidity with daily light exposures. He used a sterilized substrate of cooked wheat barries at 50% moisture. Inoculum was obtained from stipe tissue or ascospore culture. After the growth of mycelium, axenic conditions were not maintained. Ascocarp development was preceded by vegetation hyphae, conidia and sclerotia.

Subsequently two patents describing a process for the cultivation of morels have appeared in USA (Ower *et al.*, 1986 and Ower *et al.*, 1988). Royse (1995) commenting on morel cultivation maintains that these patents have revealed some of the processes involved in predictable production of sporocarp, but attempts to practice the invention have met with limited success. The commercial production of morels (in USA only) is a relatively recent development and only one company is producing morels on a commercial scale. Royse (1995) further provided the following details of the process under trial:

"Commercial cultivation involves mainly the production of sclerotia and early over-wintering stage of mushroom. Briefly 'nutrient primed' sclerotia are produced in soil placed on a layer of sterilized wheat. The production of sclerotia requires about 18-24 days under ideal conditions. At maturity the sclerotia are harvested, soaked in clean water for 24 hours and distributed into a thin layer of pasteurized 'nutrient poor' soil mix. The sclerotia germinate via the production of mycelium. After the mycelium has spread throughout the soil, a continuous (24-hour duration) fine mist of clean water is provided to stimulate the formation of ascocarps."

At present, there are five partners (Neogen, Domino's, Skandigen of Sewden, Kuhn Champignon of Switzerland and the Salk Institute Biotechnology Associates of La Jolla, CA) in the 'Company M'. The company has built pilot plant in Michigan and is expected to sell morels in the limited markets very soon. Many people still believe that Company M cannot grow morels. However, Volk (1990) has categorically mentioned that he visited their growth chambers and witnessed approximately fifty morels emerging from each plant containing soil and sclerotia. As such, there should not be any doubt that morels cannot be domesticated.

A number of problems still remain to be solved before it is accepted as a satisfactory commercial process. Consistent fruiting, competitive weed moulds, poor yield and small size are some of these problems seeking immediate attention. Further research is still needed to understand the environmental and genetic factors involved in the regulation of fruiting.

CHAPTER 3

Systematic Treatment of Morels–I

CHAPTER OUTLINE

- Introduction
- Review of Literature
- Review of Work Done in India
- Systematic Treatment of Species of *Morchella*
- Key to the Indian Species of *Morchella*
- Discussion

INTRODUCTION

There has been little consensus on the systematic status of the species comprising the genus *Morchella*. Some recognize only 3-5 species while others as many as 50 or more (Rifai, 1968; Kimbrough, 1970; Korf, 1973; Smith and Weber, 1983; Jacquetant, 1984). This problem has stemmed from the fact that there is lack of detailed study of different developmental stages of the various species and lack of data on mating and reproduction (Gessner *et al.*, 1987). Although there have been a number of studies dealing with culture, spore germination and physiology of morels (Heim, 1936; Brock, 1951; Robbins and Hervey, 1959, 1965; Gilbert, 1960; Cailleux, 1969; Delmas and Poitou, 1974; Hervey *et al.*, 1978), only recently has been some success achieved in studying life cycle in the laboratory (Ower, 1982) or understanding natural propagation in the wild (Schmidt, 1983).

In India various attempts have been made periodically (Batra and Batra, 1963; Sohi et al., 1965; Kaul, 1976, 1981; Waraitch, 1976; Wakode, 1983 and Kamal, 2005 and Kaviyasaran *et al.*, 2006) by researchers to collect and describe the representative species in the genus, but all the species have never been collected simultaneously. In all these studies (except Batra and Batra, 1963), no attempt has ever been made to consider developmental stages of sporophores to see if the species described so far exhibit intergrades or overlapping characters or not.

In North-Western Himalayas, a systematic survey of mushrooms and toadstools has been carried out by Lakhanpal and co-workers ever since 1977 (Sharma, 1980; Sharma and Lakhanpal, 1981; Bhatt, 1986; Sharma, 1986; Kaisth, 1985; Kumar, 1987; Lakhanpal, 1987; Sharma, 1987; Bhatt and Lakhanpal, 1988). Work on morels was also initiated around 1981. The following aspects were covered during the survey work on morels:

(i) To systematically collect morel species from different forest types and altitudes.

(ii) To identify and describe the collected species.

(iii) To make a comparative study of the species considering their different developmental stages, and to understand and clarify the species concept.

(iv) To fix fruit bodies at initial stages of development for cytological/developmental studies.

REVIEW OF LITERATURE

The genus *Morchella* is one of the oldest genera of Pezizales, established as early as 1719 by Dillenius. The limits of its existence, though there has been some difference of opinion with regard to the placement of species, having base of pileus free from stipe. Saccardo (1889); Seaver (1928, 1942); Imai (1932, 1954); Moser (1951); Groves and Hoare (1953); Bresinsky and Stangal (1961); Eckblad (1968); Rifai (1968) include such species in the genus *Morchella*, whereas others like Dennis (1960, 1968) and Moser (1963) treated these in the genus *Mitrophora* Lev. and still continue to treat them similarly. Eckblad (1968) in his comments on the genus *Morchella* emphatically suggested that "the genus *Mitrophora* Lev. should undoubtedly be united with it". However, Moser (1951) first included *Mitrophora* Lev. in *Morchella* but later on (Moser, 1963) separated it from *Morchella* and treated it as a separate genus.

The species differentiation in the genus is primarily based on morphological characters which are not always absolute. This resulted in the recognition of a large number of species from time to time. For example, Boudier (1907) recognized as many as 32 species. Seaver (1928) brought down the number to six, viz. *M. crassipes* (Vent.) Pers.; *M. esculenta* (L) Pers.; *M. conica* Pers.; *M. deliciosa* Fries; *M. angusticeps* Peck; *M. hybrida** (Sow) Pers. This differentiation of species was based upon whether the pileus was adnate or free, colour of the ribs and pits, depth of the pits, size and shape of the pileus and stipe. Groves and Hoare (1953) in their studies on the Helvellaceae of the Ottawa district (Canada) described three species, viz. *M. crassipes* (Vent.) Pers.; *M. angusticeps* Peck. and *M. hybrida** (Sow) Pers. They concluded from detailed field observations that *M. deliciosa* Fries and *M. esculenta* (L) Pers. are only different stages in the development of the fruit body of *M. crassipes* (Vent.) Pers. and *M. deliciosa* Fries being the earliest stage. They tagged the fruit bodies of *M. deliciosa* Fries in the field. After certain period of growth, they observed that these fruiting bodies of *M. deliciosa* Fries became identical with that of *M. esculenta* (L) Pers., and finally with that of *M. crassipes* (Vent.) Pers.

Overholts (1934) in his studies "The morels of Pennsylvania", considered *M. deliciosa* Fries; *M. esculenta* (L) Pers.; *M. crassipes* (Vent.) Pers. as distinct species; *M. deliciosa* Fries being distinguished by its markedly smaller size. He also considered *M. conica* Pers. as only a form of *M. esculenta* (L) Pers, while, according to Seaver, it is a distinct species. Groves and Hoare (1953) are of the opinion that *M. conica* Pers. is an immature stage of *M. angusticeps* Peck. Ecblad (1968) in his monograph 'The genera of Operculate Discomycetes: A re-evaluation of their taxonomy, phylogeny and nomenclature', described three species of *Morchella* viz. *M. conica* Pers.; *M. esculenta* (L) Pers.; *M. elata* Fr. Dennis (1960, 1968) in his 'British ascomycetes' described three species of genus *Morchella*, viz. *M. esculenta* (L) Pers.; *M. elata*[1] Fr. and *M. hortensis* Boud.; the species were differentiated on the basis of colour and arrangement of the ridges, size and shape of the pileus and stipe. Miller (1981) described four species of the genus from North America, viz. *M. esculenta* (L) Pers.; *M. deliciosa* Fries; *M. angusticeps* Peck. and *M. semilibera* Fr. Lincoff (1981) recorded three species, viz. *M. elata* Fr.; *M. esculenta* (L) Pers. and *M. semilibera* Fr. from North America. He merged *M. angusticeps* Peck., and *M. conica* Pers. in *M. elata* Fr. and *M. crassipes* (Vent.) Pers. and *M. deliciosa* Fries in *M. esculenta* (L) Pers. Pacioni (1980) described seven species of *Morchella*, viz. *M. semilibera* Fr., *M. angusticeps* Peck.; *M. semilibera crassipes; M. elata* Fr.; *M. conica* Pers.; *M. esculenta* (L) Pers.; *M. umbrina, M. vulgaris*[2] Smith.

Smith *et al.*,(1983) described six species of *Morchella*, namely *M. angusticeps* Peck.; *M. crassipes* (Vent.) Pers.; *M. deliciosa* Fries; *M. semilibera* Fr. and *M. esculenta* (L) Pers. Weber and Smith (1985) described four species from Southern America viz. *M. semilibera* Fr.; *M. elata* Fr.; *M. crassipes* (Vent.) Pers. and *M. deliciosa* Fries.

Mahu (1987) reported *M. vulgaris* Pers. ex Fr. for the first time from Chile. He also reported *M. conica* var *angusticeps* (Peck.) Speg. Zang (1987) described *M. tibelica* Zang, a new species in the genus *Morchella*, from China which is closely related to *M. angusticeps* Peck. This species is stated to have stipe with rough surface.

The correct name *M. hybrida** (Sow) Pers. is *M. semilibera* DC ex Fr. (Rifai, 1968) and the same has been used in this text even though Seaver's concept has been followed in the treatment of species.

[1]*M. elata* Fr. is *M. angusticeps* Peck, sensu Seaver (1928, 1942).

[2]*M. vulgaris* Smith is treated as *M. esculenta* by Seaver (1928, 1942).

REVIEW OF WORK DONE IN INDIA

The earliest report of *Morchella* from India is by Cooke (1870) as *M. gigaspora* Cooke and *M. deliciosa* Fries in 'Kashmir Morels'. Later on Cooke (1876) reported *M. esculenta* (L) Pers. Hennings (1901) in 'Fungi Indiae Orientalis II' reported *M. conica* Pers. and *M. conica* var. *acuminata* Kickx in 'Fungi aliquot Bombeyebsis'. Sydow and Butler (1911) in 'Fungi Indiae Orientalis Pars III' reported two species, viz. *M. esculenta* (L) Pers. and *M. deliciosa* Fries. Butler and Bisby (1931) reported *M. bohemica* Krombh, which is now included in the genus *Verpa* as *V. bohemica* Krombh (Seaver, 1928). Bhattacharya and Baruah (1953) reported *M. vulgaris* Krombh. from Assam, but Seaver (1928) treated this as a synonym of *M. esculenta* (L) Pers. Gosh and Pathak (1962) in their compilation described five species of *Morchella* from India. The species are *M. hybrida* (Sow.) Pers., *M. angusticeps* Peck., *M. conica* Pers., and *M. deliciosa* Fr., *M. esculenta* (L) Pers. The species were differentiated on the basis of characters: pileus adnate or free from stem at the base, shape of pileus, colour of the ribs and pits. They also doubted the report of *M. vulgaris* Krombh. of Bhattacharya and Baruah (1953) from Assam forests. Batra and Batra (1963) in their compendium 'Indian Discomycetes', reported *Morchella angusticeps* Peck.; *M. esculenta* (L) Pers.; *M. conica* Pers. (sensu Seaver); *M. crassipes* (Vent.) Pers. (sensu Seaver); and *M. deliciosa* Fries (sensu Seaver) from India as well as from USA. They merged *M. conica* Pers.; *M. crassipes* (Vent.) Pers, and *M. deliciosa* Fries with *M. esculenta* (L) Pers. They concluded that there is no difference in the tissue structure of these fungi. When a large collection is available from a single locality, various stages of development can be found.

Sohi *et al.* (1965) reported *M. angusticeps* Peck.; *M. conica* Pers.; *M. esculenta* Peck. and *M. hybrida* (Sow) Pers. from Himachal Pradesh. They separated the species on the same basis as Seaver (1928, 1942). Tilak and Rao (1968) listed *M. rotunda* (Pers.) Boud. (*M. esculenta* Auct.) Waraitch (1976) described six species viz. *M. angusticeps* Peck.; *M. crassipes* (Vent.) Pers.; *M. conica* Pers.; *M. esculenta* (L) Pers.; *M. semilibera* DC. ex Fr. and *M. deliciosa* Fries from Jammu and Kashmir, four collected by him and two by others. The basis of differentiation of species was mainly the same as by Seaver (1942); i.e. base of the pileus free or adnate from stipe, colour of the ribs and pits, size and shape of the pileus and stipe.

Kaul *et al.* (1976) and Kaul (1981) made a comprehensive study of the genus *Morchella* in Jammu and Kashmir and reported five species of *Morchella*, namely *M. esculenta* (L) Pers.; *M. conica* Pers.; *M. hybrida* (Sow.) Pers.; *M. angusticeps* Peck.; and *M. deliciosa* Fries.

Kanniyanan and Ramaswamy (1980) reported that seven species of *Morchella* have been recorded in India, but listed only *M. esculenta* (L) Pers., *M. crassipes* (Vent.) Pers. and *M. conica* Pers. Purkayastha and Chandra (1976) in the compilation on 'Edible Fungi' reported five species of *Morchella* viz. *M. deliciosa* Fr.; *M. conica* Pers.; *M. hybrida* (Sow) Pers. Purkayastha and Chandra (1985) reported, *M. deliciosa* Fr.; *M. conica* Pers.; *M. angusticeps* Peck.; *M. esculenta* (L) Pers.; Boud. and *M. vulgaris* (Pers.) Boud. Out of these, six species are valid. *M. rotunda* (Pers.) Boud., and *M. vulgaris* (Pers.) Boud. are treated as synonyms of *M. esculenta* (L) Pers. by Seaver (1928, 1942).

The literature reviewed above differentiates species in *Morchella* on morphological features only. Recently, Gessner *et al.* (1987) tackled the problem of species delimitation on genetical basis. They studied allelic variation and segregation in *M. deliciosa* Frises and *M.*

esculenta (L) Pers. They observed that these species exist as Mendelian populations due to the presence of polymorphism, the possibility of mating types and production of ascospores. The occurrence of gene exchange among sexually reproducing individuals enables a biological basis of the data. They concluded that local populations of *M. deliciosa* Fries and *M. esculenta* (L) Pers. can be reproductively isolated and exist as separate gene pools.

The details of standard methodology for collection, preservation and drying, macroscopic and microscopic studies are given in Annexure – I.

SYSTEMATIC TREATMENT OF SPECIES OF *MORCHELLA*

Genus *Morchella* Dillenius ex Fries, Syst. Myc. 2:5, 1822

Apothecia terrestrial, stipitate, hollow, differentiated into pileus and stipe. Pileus conical to almost globose in outline but very variable in shape and size; hymenium covers broad depressions separated by narrow, more or less reticulated or longitudinally arranged sterile ridges; below either the pileus is delimited from stipe by a shallow furrow or the lower part of the pileus is free from stipe. The stipe is often almost as broad as the pileus or narrower, cylindrical, eight spored, non-amyloid, not protruding at maturity. Spores ellipsoid, hyaline, smooth, without oil drops, but with external droplets at each pole, adhering to a perishable, mucous membrane. The spores contain 15 to 60 nuclei at maturity. Although hyaline, the spores are cream coloured in the mass. Paraphyses, septate, straight, enlarged above.

Type species: *Phallus esculentus* L.

KEY TO THE INDIAN SPECIES OF *MORCHELLA*

1. Pileus free from the stipe at base	....	*M. semilibera*
1. Pileus adnate to the stipe at base		2
2. Ribs or ridge of the pileus white	....	*M. simlensis* sp. nov.
2. Ribs of the pileus darker or similar in colour to the pits	....	3
3. Ribs of the pileus darker than the interior of the pits, smoky brown to almost black	....	4
3. Ribs of the pileus and interior of the pits similar in colour, slightly lighter but not white	....	5
4. Surface of the stipe smooth	....	*M. angusticeps*
4. Surface of the stipe rough	....	*M. tibelica*
5. Pileus small, 2-5 cm long, 1-2 cm broad	....	*M. deliciosa*
5. Pileus large, 4-10 cm long, 1-2 cm broad	....	6
6. Stipe conspicuously swollen at base, pits large (4 × 1 cm) usually shallow, ribs thin	....	*M. crassipes*

6. Stipe not conspicuously swollen at base, almost uniform pits smaller (2 × 1 cm), ribs thick 7
7. Pileus elongated and strongly attenuated upwards, apex cute .... *M. conica*
7. Pileus ovate, apex obtuse .... *M. esculenta*

Morchella angusticeps Peck. Ann. Rep. N.Y. State Mus. 32-44:1879

Plate I, a and Fig. 3.1

Apothecia

Stipitate, hollow differentiated into pileus and stipe. **Pileus** elongated, sub-globose, or elongated with slightly subconic apex and adnate at the base, 3.0-6.0 cm long and 2.0-4.0 cm broad. **Pits** elongated to narrowish, yellowish within and smoky brown to almost black at margins. **Ribs** somewhat longitudinally disposed and irregularly anastomosed often sharp edge, black, darker than interior of pits.

Stipe stout hollow and brittle when dry, 5.0-7.0 cm long and 2.0-3.0 cm broad, white to yellowish, slightly irregularly lacunose near the base, pubescent; hairs subhyaline compacted into conical fascicles, composed of 8 to 10 cells, cells thin walled, variously shaped.

Asci cylindrical, subcylindrical, apex obtuse, eight-spored, 240-320 μm in length and 14-22 μm in diameter; **ascospores** uniseriate, eight, 12-20 × 7-10 μm, yellowish in mass, hyaline singly, ellipsoid, smooth, eguttulate but with few external polar oil drops at each end; **paraphyses** stout thin walled, septate, simple or branched below, light brown in mass, subhyaline individually, 240-320 × 6-7 μm.

Habit and Habitat

Solitary on ground covered by dead decaying organic matter, rich in coniferous humus and mixed coniferous forests of *Cedrus deodara, Pinus gerardiana, Quercus dilatata, Picea smithiana.*

Specimens Examined

H.P. Kinnaur, Nichar, April 1986, HPUB, 20001; H.P. Shimla, Narkanda, May, 1986, HPUB, 20003; H.P. Kullu, Kandugad, March, 1985, HPUB 20005; H.P. Chamba, Himgiri. Tisa, Bhandal, May, 2001, HPUB, 1001; H.P. Kinnaur, Roghi, Kalpa, Chini, Nigulsari, Sangla, Kamru, May, 2002, HPUB, 101.

Morchella conica Pers. Champ. Comest. 257:1819

Plate I, b and Fig. 3.2

Apothecia

Stipitate, hollow differentiated into stipe and pileus. **Pileus** elongated 3-6 cm long and 1-3 cm wide more or less conical in shape acute at the apex subobtuse, more or less conical in shape and adnate at the base; **Pits** elongated and more or less rectangular in form, yellowish within when young, becoming brownish or black at maturity, longitudinally arranged, connected by

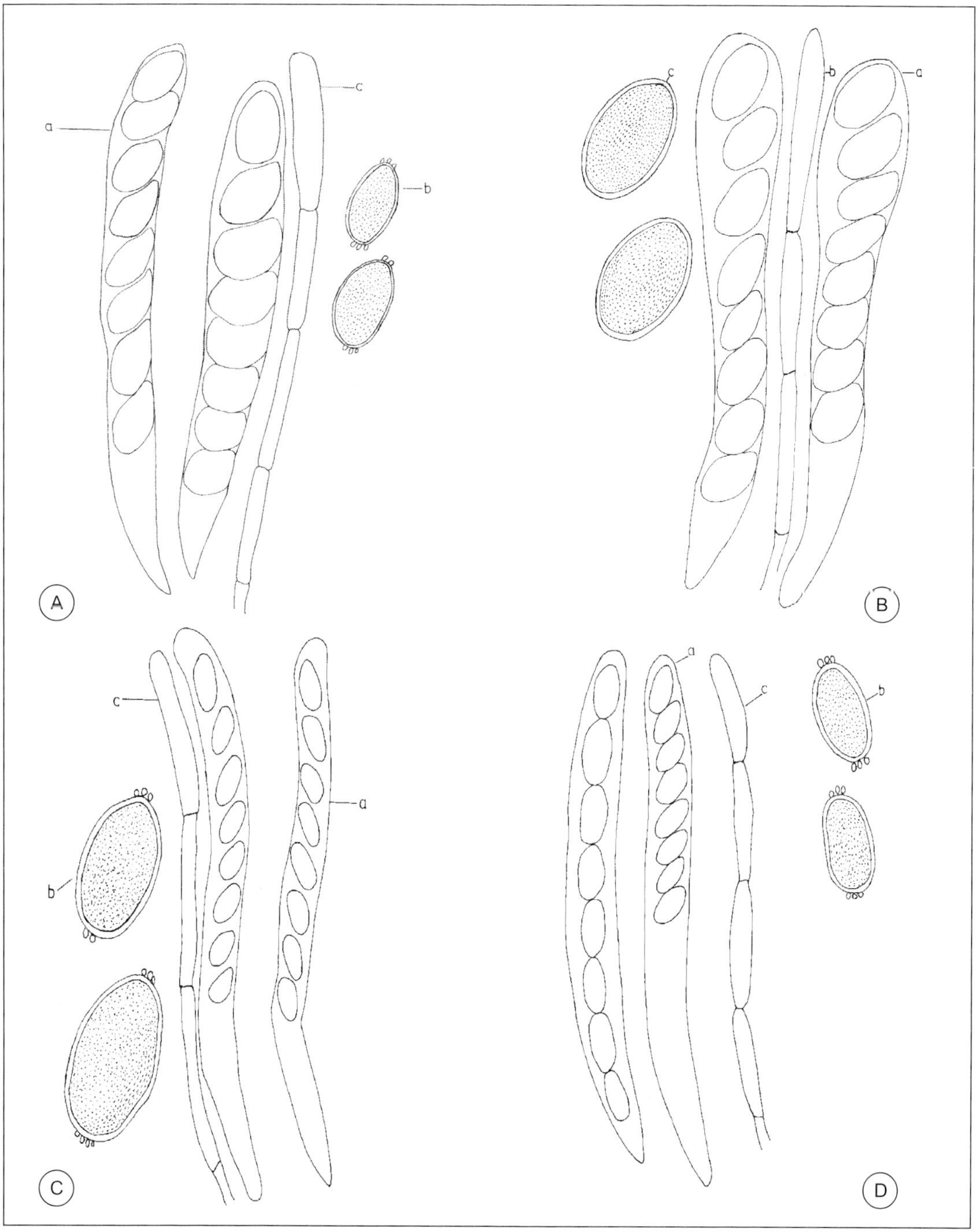

Fig. 3.1 (A–D) A – *Morchella angusticeps*; B – *Morchella conica;* C – *Morchella crassipes;* D – *Morchella rotunda*; a – Asci x 1040 b – Spores x 2350 c – Paraphyses x 1040

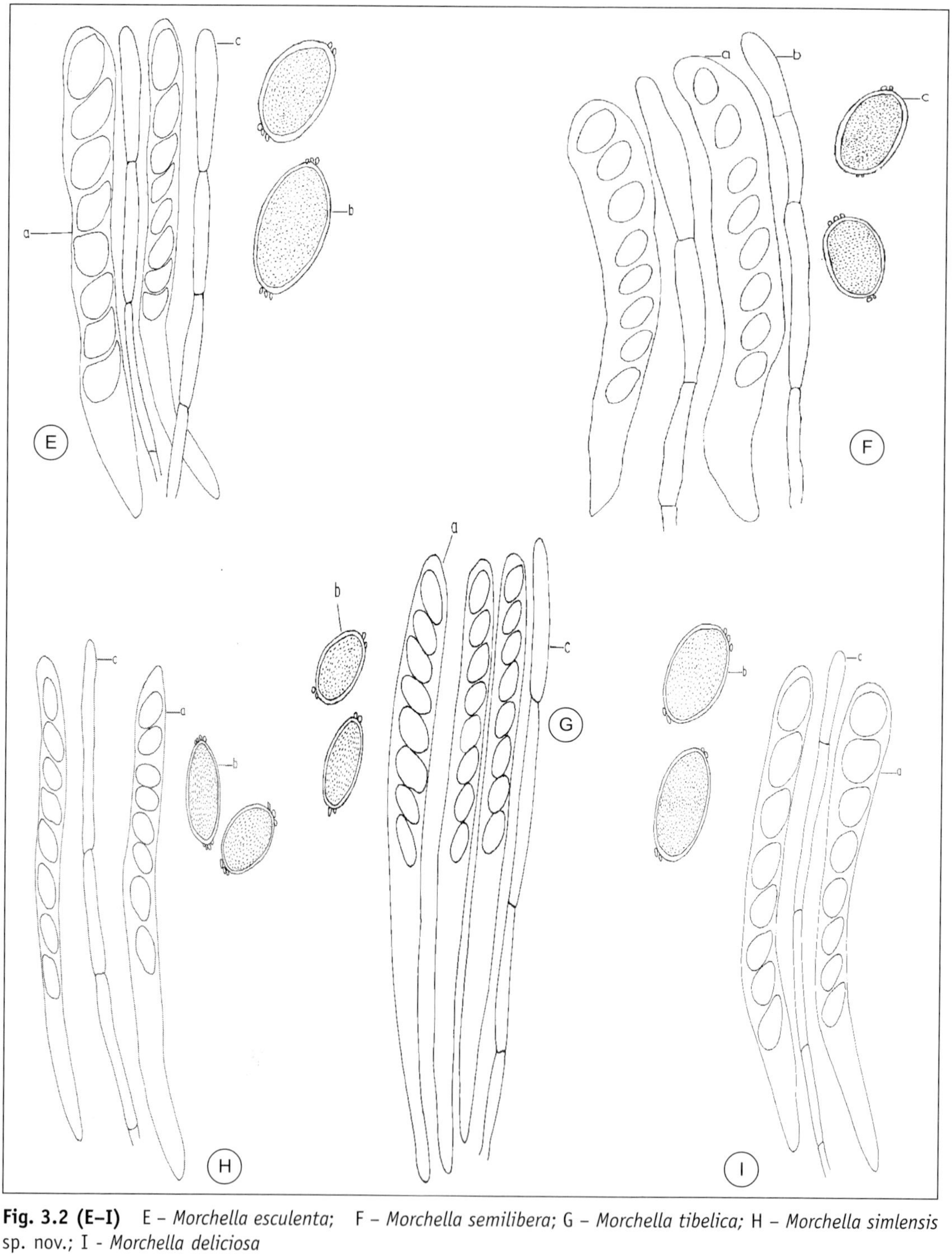

Fig. 3.2 (E–I) E – *Morchella esculenta;* F – *Morchella semilibera;* G – *Morchella tibelica;* H – *Morchella simlensis* sp. nov.; I - *Morchella deliciosa*

a – Asci x 1040 b – Spores x 2350 c – Paraphyses x 1040

PLATE I (A–I) Species of *Morchella* in Himachal Himalaya *a. Morchella angusticeps* Peck. x 3.5; *b. Morchella conica* Pers. x 3.5; *c. Morchella crassipes* (Vent.) Pers. x ½; *d. Morchella deliciosa* Fries x 1.5; *e. Morchella esculenta* (L) Pers. x ¼; *f. Morchella rotunda* (Pers.) Boud. x 1/5; *g. Morchella semilibera* DE ex Fr. x ½; *h. Morchella tibelica* Zang x 1/3; *i. Morchella simlensis* sp. nov. x 1/5

PLATE II (a–f) **(a–d)** - **Fructifications** of *Morchella esculenta* (e) *Morchella conica* (c) *Morchella deliciosa* (d) along with the developmental stages. **(e)** – A collection of morels made from one locality. **(f) - Six valid species** of ***Morchella*** placed and photographed together (*M. angusticeps, M. conica, M. crassipes, M. deliciosa, M. esculenta* and *M. semilibera*).

cross ribs, the edges rounded and lighter in colour than interior of the pits. **Stipe** stout, hollow, attenuated at the base, nearly even above, irregularly lacunose below, almost cylindrical or swollen at the base, pubescent, hairs subhyaline, composed of upto 8 cells; thin walled and variously shaped.

Asci cylindrical, 320-390 µm in length and 18-20 µm in diameter, apex obtuse, eight-spored; ascospores uniseriate, eight in number, yellowish in mass, hyaline, ellipsoid, smooth 12-21 × 10-13 µm, eguttulate but with few external polar oil drops at each end; **paraphyses** thin walled septate simple or branched below, yellowish in mass, hyaline singly, 390 - 410 µm long and 6 – 7.5 µm in diameter.

Habit and Habitat

Solitary and scattered on ground covered with dead organic matter rich in humus, it occurs in mixed coniferous forests of *Cedrus deodara, Pinus gerardiana* (chilgoza pine), *Quercus dilatata, Picea* spp., etc.

Specimens Examined

H.P. Shimla, Summer Hill, March, 1985, HPUB 20025; H.P. Kinnaur, Nankhari, April, 1985, HPUB 20030; H.P. Chamba, Himgiri. Tisa, Bhandal, May, 2001, HPUB, 1020; H.P. Kinnaur, Sangla, Chitkul, Khrogla, Roghi, Kalpa, Chini, Kashmir, May, 2002, HPUB, 140.

Morchella crassipes (Vent.) Pers. Syn. Fung. 621:1801

Plate I, c and Fig. 3.1c

Apothecia

Stipitate, hollow, differentiated into pileus and stipe. **Pileus** elongated and often subconic, adnate to the stipe at the base, 8.0-12.0 cm long and 2.0-4.0 cm in diameter, **pits** roundish or irregularly elongated, yellowish, greyish within, becoming brownish or blackish after drying, ribs and pits concolourous with minor variation at times; **ribs** irregularly anastomosed, thin, often sharp edged, edges lighter than the interior of the pits, i.e. whitish or yellowish.

Stipe stout, hollow, cylindrical 6.0-10.0 cm long and 2.0-4.0 cm broad and strongly swollen near the base, dirty white to yellowish in colour. Irregularly lacunose at the base, pubescent, hair subhyaline, compacted into conical fascicles, composed of loosely, jointed sub-globose to globose cells which are thin walled.

Asci cylindric or subcylindric, 200-2600 µm in length and 12- 20 µm in diameter, eight-spored, obtuse apex; **ascospores** eight, uniseriate, ellipsoid, 15 – 20 × 8 12 µm, yellowish in mass, hyaline or subhyaline singly, smooth, eguttulate but with external polar oil drops at each end; **paraphyses** stout, thin walled septate, simple or branched below, upto 300 µm long and 5–9 µm wide, yellowish in mass hyaline singly, enlarged above.

Habit and Habitat

Solitary or scattered, on ground covered with dead organic matter rich in humus, it occurs in mixed coniferous forests of *Cedrus deodara, Pinus gerardiana* (chilgoza pine), *Quercus dilatata, Taxus, Picea* spp.

Specimens Examined

H.P. Shimla, Nankheri, April 1986, HPUB, 20047; Summer Hill, March, 1986, HPUB, 20043; H.P. Kullu, Jalori Pass, May, 1986, HPUB 20048; H.P. Chamba, Bani, April, 2001, HPUB, 1040; H.P. Kinnaur, Sangala, Nigulsari, Roghi, Chitkul, Kamru, Rachham, May, 2002, HPUB, 160.

Morchella deliciosa Fr., Syst. Myc. 2:8 -1822

Plate I, d and Fig. 3.2I

Apothecia

Stipitate, hollow differentiated into pileus and stipe. **Pileus** elongated to sub-globose, obtuse at the apex, adnate to the stipe at base, 2.0-3.0 cm long and 1.0-2.0 cm broad, **Pits** whitish to yellowish within when fresh, becoming brown after drying, ribs and pits concolourous with some minor variations; Ribs longitudinally disposed and irregularly anastomosed, the edges rounded, slightly lighter than the interior of the pits.

Stipe stout, hollow and 1.5-2.0 cm in length and 1.0-2.0 cm wide, irregularly lacunose, white to yellowish in colour pubescent, hairs subhyaline, composed of upto 8 cells thin walled and of various shapes.

Asci broad, eight spored, cylindrical to subcylindrical, apex obtuse; ascospores eight, uniseriate, ellipsoid, yellowish in mass, hyaline singly, 10-20 μm long and 8-10 μm broad, smooth egullutate but with few polar I drops at each end; Paraphyses upto 260 μm long and 6-7 μm broad, stout, thin walled, septate, simple, light brown in mass subhyaline individually.

Habit and Habitat

Solitary or scattered on ground covered with dead organic matter rich in humus. Occurs in coniferous forests and in mixed coniferous forests of *Cedrus deodara, Pinus gerardiana* (chilgoza pine), *Quercus dilatata*, etc.

Specimens Examined

H.P. Kullu, Bahu, March, 1986, HPUB 20055; Summer Hill, March, 1986, HPUB 20059; H.P. Chamba, Dalhousie, Salooni, April, 2001, HPUB, 1061; H.P. Kinnaur, Nichar, Rally, Kalpa, Rackham, Dangelanshi, Kamru, April, 2002, HPUB, 180.

Morchella esculenta (L) Pers. Syn. Fung. 618-1801

Plate I, e and Fig. 3.2.E

Apothecia

Stipitate, hollow differentiated into stipe and pileus. **Pileus** sub-globose ovate or elongated, Attenuated upwards but obtuse at the apex and pileus adnate to the stipe at the base, varying much in size, but often reaching a length 4.0-10.0 cm and diameter of 2.5-5.0 cm, **pits** colour yellowish within when fresh becoming brown after drying; ribs and pits are concolourous with minor variation at times; ribs irregularly anastomosed, not longitudinally disposed, the edges rounded and lighter than interior of the pits, i.e. usually whitish or yellowish.

Stipe stout, hollow, usually little enlarged at the base and nearly even above. Colour whitish to yellowish, 4.0-8.0 cm × 2.5-5.0 cm delicately pubescent, brittle when dry, hairs subhyaline, composed upto 8 cells, cells thin walled.

Asci usually 200-350 μm long and 12-23 μm broad, eight spored, narrowed below; **Ascospores** eight, uniseriate, 15-23 μm × 8-13 μm yellowish in mass, hyaline singly, ellipsoid, smooth, eguttulate, but with external polar oil drops at each end; **paraphyses** stout thin walled septate simple, branched below, yellowish in mass, hyaline individually enlarged above, 4-8 μm broad and 32-340 μm long.

Habit and Habitat

Solitary or scattered on ground covered with dead organic matter rich in humus. It occurs in coniferous and mixed coniferous forests of *Cedrus deodara, Pinus gerardiana* (chilgoza/Neoza pine), *Quercus dilatata* (Oak), etc.

Specimen Examined

H.P. Chamba, Tisa, April, 1986, HPUB 20085; H.P. Kullu, Pugla, April, 1985, HPUB 20068; Chowai, March 1985, HPUB 20096; H.P. Shimla, Narkanda, May 1986, HPUB 20105; H.P. Hamirpur, Jolsapar, October, 2001, HPUB 1071; H.P. Chamba, Khajjiar, Bhandal, April, 1001, HPUB, 1071; H.P. Kinnaur, Sangla, Kalpa, Chitkul, Nigulsari, Roghi, May, 2002, HPUB, 200.

Morchella semilibera DC ex Fr. Verhandel Konik, Nederl. Akad. Wetensch; Afd. Natuvrk. 57: 1-295. 1968

Plate I, g and Fig. 3.2.F

Apothecia

Stipitate, hollow differentiated into pileus and stipe. **Pileus** conic to elongated, free at the base, 2-3 cm and 1-2 cm broad, apex acute or obtuse, light brown in colour; **pits** irregularly rounded or elongated, shallow brownish or blackish; **ribs** almost longitudinal to irregular anastomosed, edges rounded, lighter than the interior of the pits.

Stipe long, hollow, brittle when dry, 8-10 cm long, tapering upwards, yellowish white, irregularly lacunose at the base, nearly even above delicately pubescent, hairs subhyaline, composed of 8–10 cells, cells thin walled, variously shaped.

Asci cylindrical, 210–260 μm long and 12-19 μm broad, eight spored, apex obtuse; **ascospores** light and uniseriate, ellipsoid, yellowish in mass, hyaline singly, smooth, eguttulate but with external polar oil drops at each end, **paraphyses** stout, thin walled septate, simple or branched below, light brown in mass subhyaline individually, 200–250 μm.

Habit and Habitat

Solitary and scattered on ground covered with dead decaying organic matter in humus in silver fir–spruce forests.

Specimens Examined

H.P. Shimla Chazpur, April 1986, HPUB 20144; Chazpur May, 1985, HPUB 20148.

Morchella tibelica Zang. Acta. Bot. Yunnanica a(1): 81-88:1987

Plate I, h and Fig. 3.2.G

Apothecia

Stipitate, hollow differentiated into pileus and stipe. **Pileus** elongated and conical, apex conic and base adnate to stipe, 4.0-7.0 cm long and 2.0-4.0 cm broad; **pits** are elongated, yellowish to brownish within and brown at the margins, pits are not concolourous with ribs; **Ribs** longitudinally disposed and irregularly anastomosed, sharp edged, black, darker than the interior of pits.

Stipe stout, hollow and brittle when dry, rough surfaced, white and yellowish, irregularly lacunose near the base, 2.0-4.0 cm long and 1.0-2.0 cm in diameter, pubescent, hairs subhyaline, cells thin walled, composed of eight cells which are variously shaped, surface rough.

Asci cylindrical or subcylindrical, apex obtuse and eight-spored; **ascospores** eight, uniseriate, yellowish in mass, hyaline singly ellipsoid, smooth eguttulate, but with few external oil drops at each end; **paraphyses** stout, thin walled, septate, simple or branched below, light brown in mass subhyaline individually, 210–305 µm and 5-6 µm.

Habit and Habitat

Solitary on ground covered with dead decaying organic matter rich in humus in coniferous, mixed coniferous and oak forests.

Collection Examined

H.P. Kullu, Khanag, March, 1985, HPUB, 20166; H.P. Kinnaur, Nichar, April 1985, HPUB, 20167; H.P. Shimla, Nankheri, April, 1986, HPUB 20178; H.P. Kinnaur, Nichar, Nigulsari, Powari, Kamru, April, 2002, HPUB, 250.

Morchella simlensis Lakhanpal & Shad sp. nov.

Plate I, i and Fig. 3.2.H

Apothecium 6.0-11 cm alatum, pileus globosus vel elongatus, 4-6 cm latus, stipes 2-5 cm longus, pits brunneae, costae albus, sporae 12-26 × 10-15 µm. Holotypus; HPUB 220, H.P. Kinnaur, Sangla, Chitkul, April-May, 2002

Apothecia stipitate, hollow differentiated into pileus and stipe. **Pileus** globose or elongated, obtuse apex, adnate to the base, 4.0-6.0 cm in diameter; **pits** are brownish to yellowish in colour from within when fresh. They turn to brown colour after drying. **Ribs** are irregularly anastomosed with sharp edges. They are clear white in colour, much lighter than the interior of the pits.

Stipe stout, hollow and brittle when dry, irregularly lacunose nearly even above, whitish to yellowish in colour. 2.0-5.0 cm long and 2.0 cm in diameter, delicately pubescent, hairs subhyaline, composed of eight cells, cells are thin walled and variously shaped.

Asci cylindrical to subcylindrical, apex obtuse, eight-spored; **ascospores** light and uniseriate, 12-26 × 10-15 µm, ellipsoid, smooth, eguttulate but with internal oil drops at each end, yellowish in mass and hyaline singly; **paraphyses** individually, enlarged above 320-350 µm long and 4-8 µm in diameter.

Habit and Habitat

Solitary on ground covered with dead organic matter in chir–oak and mixed coniferous forests.

Specimens Examined

H.P. Shimla, Summer Hill, March, 1986, HPUB 20185; H.P. Kullu, Chowai, April, 1985, HPUB 20 188; H.P. Kinnaur, Sangla, Chitkul, April and mid-May, 2002, HPUB, 220.

The justification for treatment of these collections as a new species is given in the following discussion and in the next chapter, i.e. Chapter 4 wherein the genotypic polymorphism and diversity in morel species has been given. These populations have clear white ridges.

DISCUSSION

The systematics of the genus *Morchella* and the species contained therein have been in a state of flux as the opinions have greatly differed with regard to the number of species in the genus. Some people have recognized only three species, others six and still others as many as upto 50 or so. The confusion in species concept has stemmed from the fact that there is lack firstly, of detailed study of different developmental stages of the various species and secondly of data on mating and reproduction (Gessner *et al.*, 1987). Many workers have described species of *Morchella* from India from time to time (Batra and Batra, 1963; Sohi *et al.*, 1965; Kaul *et al.*, 1976; Kaul, 1981; Waraitch, 1976; Wakode, 1983) but all the six species have never been collected by any single worker except Lakhanpal and Shad (1987, a, b). In all these studies (except, Batra and Batra, 1963 and Lakhanpal and Shad, 1987) no mention has been made if developmental stages have ever been collected to see stability of morphological characters used in differentiating the species.

A systematic survey of morels of Himachal Himalayas (1800 to 2700 m) has been carried out by the authors from 1980 onwards. All the six traditional species, viz. *M. angusticeps* Peck.; *M. esculenta* (L) Pers.; *M. crassipes* (Vent.) Pers.; *M. deliciosa* Fr.; *M. conica* Pers. and *M. semilibera* Fr. are represented in these collections. Attempt has also been made to collect sporophores of all these species at various developmental stages to study intermediates, if any. Such collections have been possible in all the species except *M. semilibera,* which was available only in mature stage. All other species are represented by small to medium and mature spo.. ocarps.

Presently though most of the authors favour a six species concept in the genus *Morchella* (Seaver, 1928; Ecblad, 1968; Smith and Weber, 1980), even for these species there is no unanimity (Plate – II e). On the basis of the criteria used by Seaver (1942) for distributing the six species of *Morchella*, two species, i.e. *M. semilibera* and *M. angusticeps* stand out distinctly whereas others pose problems occasionally in establishing their clear cut identity, for in them the key features overlap at times. In spite of this difficulty, these basis of demarcation are till today the best for identifying and placing different species.

A study of mature sporocarps of all these species from our collections revealed that some specimens were typical and easily identifiable, whereas others, especially in the species, viz. *M. deliciosa, M. esculenta, M. crassipes* and *M. conica* were represented by intermediate forms (Plate II a - d), supporting Batra and Batra's (1963) contention that when a large collection is available from a single locality or various stages of development can be traced, specimens representing each of these species can be found. In our collections also *M. semilibera* and *M. angusticeps* can be clearly distinct, the former by virtue of its half free pileus and the latter by its dark coloured ribs. *M. conica* is also fairly distinct in some specimens, only if its shape is given prime consideration but on the basis of size, it does not separate out without some overlaps. We have specimens which fit into *M. deliciosa* and/or *M. esculenta* fairly well, but they otherwise have typically conical pileus of *M. conica*. Further in these species, some specimens have swollen stipes, resembling those of *M. crassipes* (Plate II a - d). In other words a great variation is obvious in the distinguishing features of *M. deliciosa, M. esculenta, M. crassipes* and *M. conica*. These may be taxonomically good species when described from a few isolated specimens, but certainly pose problems of appropriate placement when represented by larger populations. In such instances they all seem to be just one complex morphological species.

Due recognition was given to immature and smaller fructification during the present work. A study of the younger stages of the five species of *Morchella,* viz. *M. conica, M. angusticeps, M. deliciosa, M. crassipes,* and *M. esculenta,* revealed that it was not only difficult but impossible to distinguish just emerging fruiting body initials by themselves, if not accompanied by mature sporophores, because they all looked alike. It was difficult to place them in different species; the distinction in them became apparent only when they became medium sized. Such initials, however, usually accompanied mature fructifications of specific species, and hence this indicated, of which *Morchella* species developmental stages these were otherwise, the initial stages of all the five species were similarlv shaped and structured and not at all distinguishable. So, these observations clearly point out that not only *M. conica, M. deliciosa, M. crassipes,* and *M. esculenta* but *M. angusticeps* also is similar to them in the initial or primordial stages of development, further complicating the species concept in *Morchella.*

Lakhanpal and Shad (1987) recorded how distinctly colour in different species of *Morchella* varied with different forest types and altitudes. No wonder then that such colour variants have been given different names from time to time by various workers. Such morphological and geographic types have also been accorded specific ranks by some, but all these now need to be investigated critically in the light of Gessner *et al.*'s (1987) recent genetical segregation of *M. deliciosa* and *M. esculenta,* before discarding them altogether as mere colour variants. Hence, some of the species now reduced to synonymy or described as new deserve investigation on Gessner *et al.*'s (1987) (1. c.) lines.

Recently, Zang (1987) described *M. tibelica* Zang, a new species in the genus *Morchella* from China. This species is stated to be a close relative of *M. angusticeps,* differing from it in having a rough surfaced stipe. Our specimens of *M. angusticeps* from Himachal Pradesh are comprised of two types of populations: in one type, the fructifications are comparatively smaller and stipe smooth; in the second type, the fruit bodies are taller and stipe with rough surface. The latter populations, therefore, represent what has been recently described by Zang as *M. tibelica* and, the former is typical *M. angusticeps.* It has been noticed that rough stipe is characteristic of taller fructifications and smooth of smaller ones always, to be merely coincidental and, therefore, these certainly represent two distinct groups of specimens in the same or in different species, as has been done by Zang (1987). But whether the rough stiped, taller populations really need a separate name remains to be proven by genetical studies. Presently, it seems best to treat them distinct from *M. angusticeps* till such studies prove them otherwise.

If these above populations in *M. angusticeps* complex are treated as two distinct species, some specimens which we have and which are characterized by clear white ridges/ribs, not recorded in any of the other species, also deserve some place in this discussion. These specimens, represented by some good populations, possess clear–white ridges. Miller (1981) included such specimens in *M. deliciosa*. But on the corollary of *M. tibelica,* these also need to be treated separately till mating behaviour is worked out or molecular studies conducted. Similarly, it is felt that *M. rotunda,* which is now treated as a synonym of *M. esculenta,* also needs reinvestigation and should be presently treated a distinct species by virtue of its typically round shape by the same parameters as *M. conica*. In our collections we have enough specimens of *Morchella* which we had earlier placed in *M. deliciosa* and/or *M. esculenta,* but which possess a typically rounded pileus apex, as has been reported for *M. rotunda,* in the literature. Hence, such specimens should also be treated separately as *M. rotunda*. A similar treatment seems necessary for *M. umbrina* (c.f. Pacioni, 1980). Even if such specimens represent mere colour variants, it is worthwhile to investigate and observe their mating behaviour (this species, is not represented in collections from Himachal Pradesh, however).

It emerges from the above discussion that in spite of all overlaps and intergrades, not only the six species, but 3-4 more seem to be distinct so long as they are described from mature fruiting bodies. The data presented on intermediate stages by Groves and Hoare (1953), Batra and Batra (1963) and our own observations suggest only three species to be clearly distinct morphologically, viz. *M. angusticeps, M. semilibera* and *M. esculenta.* Nevertheless, the satisfactory resolution of this issue will emerge only from the data on mating and reproduction in all the species of *Morchella* (Gessner *et al.,* 1987) or evaluating genetic diversity through molecular markers. Cultural studies of Harvey *et al.* (1978) of single spore ascospore isolates of *M. esculenta* suggested that mating types do exist and sexual reproduction is involved in the genus *Morchella.* Recently, Gessner *et al.* (1987) presented electrophoretic data that serves to confirm that allelic variations exist and segregate in natural populations of individuals of *Morchella,* indicating the possibility of sexual reproduction. They suggest that presence of polymorphism, the possibility of mating types, and production of ascospores indicate that *M. deliciosa* and *M. esculenta* exist as Mendelian populations. Their data further suggests that these two species are reproductively isolated, and exist as separate gene pools or distinct species. On the basis of these findings, Gessner *et al.* (1987) do not subscribe to Groves and Hoare's (1953) assumption of treating *M. deliciosa* an immature form of *M. esculenta* but maintain them as two separate taxa.

Therefore, on the basis of morphological features, out of the four species in the *M. esculenta* complex, (*M. esculenta, M. deliciosa, M. conica* and *M. crassipes*), *M. conica,* and *M. crassipes* have at least one distinguishing characteristic each (conical shape in the former and swollen stipe in the latter) by which they can be easily identified and placed in the taxonomic hierarchy. But there is hardly any such definitive feature that separates *M. esculenta* and *M. deliciosa,* so that these have been often regarded as a single species. However, the data of

Gessner *et al.* (1981) on allelic variation and segregation in these two species clearly points that these two are distinct taxa. This data plus data of Hervey *et al.* (1978) suggests that the potential for gene exchange may be used as a means of discriminating taxa within the genus *Morchella*. It emerges, therefore, that similar types of studies are needed on other species of *Morchella;* not only on the six distinct species, but also on other variants, e.g. *M. rotunda, M. tibelica, M. umbrina* and *Morchella simlensis*. sp. nov., before they are merged into the synonymy of other related species. Because of these observations it is pertinent to treat the white ribbed specimens as a new species; relegate the rough stipes longer fruiting bodies in *M. angusticeps* to *M. tibelica* and to place the round-apexed fructifications in *M. rotunda*. Till then it is better not to merge them on morphological considerations alone. As pointed out by Gessner *et al.* (1987), "definite taxonomic conclusions will require further sampling from geographically distant populations of all these species".

In the next chapter are presented results on analysis of polymorphisms in morels considering majority of the points raised in the discussion above.

CHAPTER 4

Systematic Treatment of Morels–II: Genotypic Polymorphism and Diversity

CHAPTER OUTLINE

- Introduction and Review of Literature
- Observations
- Similarity Analysis and Clustering of Data
- Discussion

INTRODUCTION AND REVIEW OF LITERATURE

The problems in the delimitation of the species of *Morchella* have been discussed in the previous chapter. To resolve some of the taxonomic problems, molecular techniques have been applied and the results of these studies are presented in this chapter. The genus *Morchella* is polymorphic in nature. It has been suggested that high degree of polymorphism might be due to the influence of environmental and climatic factors. Several genetic studies have been attempted to resolve the taxonomic problems or polymorphism in different members of Morchellaceae. The polymorphism studies were consistent with the phylogenetic relationships between the investigated strains only at genus level.

Presently, most of the authors favour a six species concept in the genus *Morchella* (Seaver, 1928, 1942; Eckblad, 1968 and Smith and Weber, 1980). On the basis of the criteria used by Seaver (1942), out of the six species, two, i.e. *M. semilibera* and *M. angusticeps* stand out distinctly whereas others pose problems occasionally in establishing their clear-cut identity due to their overlapping features. (Table 4.1)

Table 4.1 Collected populations, species, geographical sources and year of isolation of different species of *Morchella*

Population No.	*Species*	*Tissue*	*Geographical location*	*Locality*	*Year of collection*
1.	***Morchella angusticeps***	Ascocarp	India (H.P.) District Kinnaur	Sangla	2001
2.	***M. conica***	Ascocarp	-	Chitkul	2001
3.	***M. crassipes***	Ascocarp	-	Kalpa	2002
4.	***M. deliciosa***	Ascocarp	-	Nichar	2002
5.	***M. esculenta***	Ascocarp	-	Nigulsari	2001
6.	***M. tibelica***	Ascocarp	-	Kalpa	2003
7.	***M. simlensis*** sp. nov.	Ascocarp	-	Kalpa	2003

The satisfactory resolution of this issue in the species of *Morchella* was attempted by Gessner *et al.* (1987). Cultural studies of Hervey *et al.* (1978) of single ascospore isolates of *M. esculenta* suggested that mating types do exist and sexual reproduction is involved in the genus *Morchella.* Gessner *et al.* (1987) presented electrophoretic data that serves to confirm that allelic variations exist and segregate in natural populations of individuals of *Morchella,* indicating the possibility of sexual reproduction. They suggest that presence of polymorphism, the possibility of mating types, and production of ascospores indicated that *M. deliciosa* and *M. esculenta* exist as Mendelian populations. Their data further suggest that these two species are reproductively isolated, and exist as separate gene pools or distinct species. On the basis of these findings Gessner *et al.* (1987) do not subscribe to Groves and Hoare's (1953) assumption of treating *M. deliciosa* an immature form of *M. esculenta* but maintain them as two separate taxa.

Therefore, on the basis of morphological features, out of the four species in *M. esculenta* complex, (*M. esculenta, M. deliciosa, M. conica* and *M. crassipes*), *M. conica* and *M. crassipes* have at least one distinguishing characteristic each (conical shape in the former and swollen stipe in the latter) by which they can be easily identified and placed in the taxonomic hierarchy. But there is hardly any such definitive feature that separates *M. esculenta* and *M. deliciosa*, so that these have been more often regarded as a single species. However, the data of Gessner *et al.* (1987) on allelic variation and segregation in these two species clearly points that these two are distinct taxa. This data plus data of Hervey *et al.* (1978) suggests that the potential for gene exchange may be used as a means for discrimination of species within the genus *Morchella*. Similar type of studies are needed on other species of *Morchella* as well before they are merged into the synonymy of other related species. The present studies have been undertaken to genetically analyze the controversial species.

In the present-day scenario, the detection of genetic variability/diversity in different biological materials (mushrooms) has undergone transformation with the advent of molecular techniques. The ascocarps of *Morchella* are highly polymorphic in appearance, exhibiting many shapes, sizes and colour variations. The polymorphic nature of the genus has also contributed to taxonomic uncertainties within *Morchella* (Bunyard *et al.*, 1994). Morphological markers to study polymorphism in the genus *Morchella* are limited in number. Differentiation on the basis of morphological characters is also inefficient, slow and time-consuming. On the contrary, molecular markers are free from environmental effect and are innumerable in number providing better option in identifying, characterizing and generating information on polymorphism. Restriction Fragment Length Polymorphism (RFLP), Random Amplified Polymorphic DNA (RAPD), Amplified Fragment Length Polymorphism (AFLP), Simple Sequence Repeat (SSR), etc. have been used to study genetic variability within and between species. Among them the RAPD technique is being favoured largely because of their simplicity. The systematic status of controversial species of *Morchella* can be best ascertained by using the molecular techniques. In the present study, the polymorphism in different species of *Morchella* has been assessed through Random Amplified Polymorphic DNA (RAPD) markers.*

Molecular Markers:

Many biochemical compounds (like phenolics, anthocyanins, flavonoids, glycosides, etc.) have the potential to be utilized as molecular markers. However, the chromatographic techniques used to analyze these molecules are often expensive and time-consuming making them unsuitable for high through routine analysis. However, proteins have the limitation of profiling, so to overcome this limitation polymorphism in the structure of nucleic acids (particularly in DNA) is now being exploited. Deoxyribonucleic acid is the universal biomolecule, which acts as the repository of information for existence and evolution of living world. Though this message is perpetuated more or less faithfully during DNA replication, variations in base sequence do occur as a result of base substitution, deletion or insertion, transposition, gene inversion, unequal exchange during meiotic or mitotic recombination and polymerase slippage at the replication fork, giving rise to sequence polymorphism. DNA fingerprinting techniques exploit this sequence polymorphism to discriminate an individual from others.

Among the DNA-based markers, Restriction Fragment Length Polymorphism (RFLP) and Random Amplified Polymorphic DNA (RAPD) are mostly used to study fungal genetics. For RFLP and RAPD analysis, a variety of reliable methods for extracting fungal DNA (Taylor and Natvig, 1987) can be used. Initially, RFLP was the usual method to study genetic divergence within a population. RFLPs are based on variation in cleavage patterns produced by restriction endonucleases. These enzymes recognize specific short sequences in the DNA and make a cut at that particular site. However, RFLP technique is time-consuming, requires the development of polymorphic DNA probes, usually involves the use of radioisotopes and requires an expensive laboratory setup, which is beyond

(Contd.)

The *Morchella* ascocarps are polymorphic in nature. Ascocarp morphology is the primary determinant of *Morchella* species. Different systematic treatments of *Morchella* spp. have been proposed by scientists since the work of Boudier (1897). Some researchers classify the genus into 3-6 species (Boudier, 1907; Overholts, 1934; Groves and Hoare, 1953; Korf, 1972; Seaver, 1942; Smith *et al.*, 1981 and Weber, 1988), Delmas and Poitou (1974) delineated 12 species, while other mycologists recognize as many as 50 or more species (Jacquetant, 1984; Kimbrough, 1970; Korf, 1973; Miller, 1972; Pacioni, 1981; Rifai, 1968 and Smith and Weber, 1980). The species delimitation in morels is still controversial (Berthet, 1964; Bresinky *et al.*, 1972 and Korf, 1973). Therefore, there is a need to analyze the genetic make-up of *Morchella* species at molecular level to resolve the issue of species delimitation.

For such type of analysis a variety of Polymerase Chain Reaction (PCR) on DNA based fingerprinting technique like Random Amplified Polymorphic DNA (RAPD) are being used in a number of laboratories for fungal taxonomy (Williams *et al.*, 1990 and Williams and Kubelik, 1991). Even in some species of *Morchella* such type of work has been carried out. Gessner *et al.* (1987) studied the allelic variations and segregation in *Morchella deliciosa* and *M. esculenta* from North America. The presence of polymorphism, the possibility of mating types and the production of ascospores indicates that *Morchella deliciosa* and *M. esculenta* exist as Mendelian populations. Their data also indicates that local populations of *Morchella deliciosa* and *M. esculenta* are reproductively isolated and exist as separate gene pools. They presented electrophoretic data that serves to confirm that allelic variations exist and segregate in

---

(Contd.)

the economic feasibility of many institutions/universities. Therefore, a variety of Polymerase Chain Reaction (PCR) based DNA based finger printing technique like RAPD (Williams *et al.*, 1990), Amplified fragment length polymorphism (AFLP) (Vos *et al.*, 1995) and inter simple sequence repeat – Polymerase Chain Reaction (Zeitkiewics *et al.*, 1994) have been developed in the last decade, which don't require prior investments in terms of sequence analysis. The RAPD analysis introduced in 1990 is the most favoured and most powerful technique with many applications for DNA fingerprinting largely because of relative ease with which the technique can be practiced in any molecular biology laboratory (Milbourne *et al.*, 1977). RAPD analysis has been used in many findings in mycology (Williams and Kubelik, 1991).

In RAPD, a single decamer primer is used to amplify regions of genomic DNA at random. Only those regions of genomic DNA, which are flanked by inverted repeats of oligonucleotide primer, will be amplified. Amplified fragments are visualized under UV light following separation on an agarose gel and staining with ethidium bromide. Polymorphism is detected as presence or absence of a band. Heterozygotes, which theoretically should give lower intensity bands, are usually not distinguishable and for this reason the RAPD marker system is essentially a dominant system.

RAPD technique has several advantages over established RFLP methodologies as follows:

- Requires small amount of genomic DNA (15-25 ng).
- Simple experimental setup requiring only a thermocycler and Agarose gel Electrophoresis assembly.
- Results are quick, efficient screening for DNA based polymorphism, the whole analysis from reaction to gel, can be completed in single working day and availability of very large number of scorable loci.
- No requirement of specific gene library, any form of clone required as probes, all one needs is one/several primers that are commercially available.
- No radioisotopes are required giving reduction in cost and increase in safety.
- Involves no blotting/hybridization steps.

These advantages, together with the relative ease of the technical procedure lead to the increased use of RAPD in Genetic Divergence Analysis. Popularity of this technique is reflected in publications of large number of research papers on this topic every year.

natural populations of individuals of *Morchella,* indicating the possibility of sexual reproduction. The fertilization event in a single ascocarp leading to production of different asci within the ascocarp may involve several individuals, which are haploid.

Kulkarni and Kamerath (1989) compared different American *Morchella* species by isozyme analysis. They used five different mycelial species of *Morcheila,* viz. *M. angusticeps, M. crassipes, M. deliciosa, M. elata* and *M. esculenta*. Eight soluble enzymes from these five different *Morchella* species cultures were resolved by polyacrylamide gel electrophoresis and enzymes were visualized by specific activity staining procedures. The studies revealed that there were several common alleles in all the five *Morchella* species. There were close relationships between *M. elata* and *M. angusticeps,* whereas *M. esculenta* is closely related to *M. crassipes.* The genotype of *M. deliciosa* appears to be unique.

Later on, the population genetics and systematics of *Morchella esculenta* complex from West Central Illinois and South West Wisconsin in North America by using isozyme polymorphism was studied by Yoon *et al.* (1990). A total of 122 mono ascosporus isolates from 72 different ascocarps were studied, and the studies revealed that the strains of two different forms, however, do not cluster separately in a phenetic analysis indicating that the two phenotypes are not different taxa. Such data suggest that allelic frequencies of *M. esculenta* from different localities may vary considerably, yet the relationships among the distinct populations exist.

The immunological relationships among five strains of *M. esculenta* complex by using Enzyme Linked Immunosorbent Assay (ELISA) have been reported by Jung *et al.* (1993). All of these five strains were characterized as being tan, large and tan, or grey. Two strains of *M. semilibera* were chosen as morphological outgroups for comparison. The mycelial extracts were used as antigens to induce antisera production in rabbits. The result of the study indicates that grey, tan and large forms of *M. esculenta* complex are immunologically indistinguishable and likely co-specific, while *M. semilibera* is distinguished from the *M. esculenta* complex. Thus, the immunological distance data from this ELISA study support the morphological data that *M. semilibera* and the *M. esculenta* complex are distinctly different. The large tan forms of *Morchella* in this study have been referred to as *M. crassipes.* They suggested that large tan forms are just the forms of *M. esculenta.*

The systematic assessment of *Morchella* species using RFLP analysis of the 28S ribosomal RNA gene has been reported by Bunyard *et al.* (1994). The phylogenetic tree developed from the restriction data revealed that the black morels *(M. angusticeps, M. elata* and *M. conica*) and the yellow morels (*M. esculenta, M. crassipes* and *M. deliciosa*) are separate taxonomic groups. The third group, free capped morels were not the part of the study. So again, Bunyard *et al.* (1995) studied the inter- and intraspecific polymorphism among different putative species of *Morchella* (including *M. semilibera*) and to elucidate the relationships of *Morchella* to other genera of Morchellaceae on the basis of restriction analysis of the 28s Ribosomal RNA gene. The phylogenetic tree drawn suggested that black morels (*M. angusticeps, M. elata* and *M. conica*), yellow morels (*M. esculenta, M. crassipes* and *M. deliciosa*) and semi-free capped morels (*M. semilibera*) are separate, albeit, polymorphic and taxonomic groups. The black morels grouped together had much more variability than yellow species. *M. semilibera* was more distant to black *Morchella* species than yellow ones. Other two genera *Verpa* and *Disciotis* of *Morchella* can be grouped loosely. Only *Gigaspora gigas* was different from all other genera of Morchellaceae.

Later on, Buscot *et al.* (1996) studied the DNA polymorphism within morel species by PCR/RFLP analysis of the ribosomal DNA spacers and microsatellites-primed PCR. They infer that polymorphism of the internal transcribed spacer (ITS) of rDNA is more adequate to improve morel systematics. In order to assess which discrimination level is necessary to trace nucleus populations in heterokaryon, but also in the context of the debatable species definition in morels, different taxa and the strain types were analysed by two PCR techniques of distinct sensitivity. They recognize two morphological groups of species of *Morchella* (i) adnate morels (base of the hymenium is not free) to which *Morchella esculenta* and *M. crassipes* belong; (ii) the distantes species (base of hymenium is free) to which *M. conica* and *M. angusticeps* belong.

The polymorphism in morels by using isozyme electrophoretic analysis on European, American and Israelian strains has been reported by Wipf *et al.* (1996). The aim of the study was to assess the degree of isozyme polymorphism at different taxonomic levels (genus, species and intra specific) in different members of Morchellaceae and also to evaluate intraspecific crossing between monosporal strains of *M. esculenta.* For this purpose, wipfetal. performed the isozyme electrophoretic analysis of the following enzymes: glutamine synthetase, NAD-glutamate dehydrogenase, NADP-glutamate dehydrogenase, aspartate amino transferase, malate dehydrogenase, NAD-glyceraldehyde phosphate dehydrogenase, glucose phosphate isomerase, and superoxide dismutase and confirmed the findings that morels exhibit higher isozyme variations at intraspecific than at specific levels.

Again, Wipf *et al.* (1996) studied DNA polymorphism in morels by complete sequences of the Internal Transcribed Spacer (ITS) of genes coding for rRNA in *Morchella esculenta* (yellow morels) and *M. conica* (black morels). They inferred that both the groups are distinct ones. He analysed internal transcribed spacer (ITS) of rDNA by PCR/RFLP in 66 strains belonging to 11 putative species of Morchellaceae and 3 species of Discineceae. Particularly 3 species of *Morchella* viz. *M. esculenta, M. conica* and *M. elata* were isolated from different geographical origins. It was observed that no intraspecific variations of ITS length and restriction were noted and all genera were separate ones.

The genetic diversity within and between populations of morel mushrooms *M. esculenta* by using RAPD markers was observed by King and Jacobson (2000). They tested the Null hypothesis that *M. esculenta* populations in Iowa would exhibit low levels of genetic diversity using RAPD markers and infered that morel population might be highly inbred.

Singh *et al.* (2003) did the DNA fingerprinting of edible mushrooms using random primers. Mycelial cultures of nine edible mushroom strains viz. *Agaricus bisporus* (strains S-11 & U-3), *Pleurotus flabellatus, P. ostreatus, P. sajor-caju, Auricularia polytrichia, Volvariella volvacea, Lentinula edodes* and *Morchella esculenta* were raised in pure cultures, and subsequently dried, DNA was isolated of all the mushrooms mycelia and RAPD carried out. The result of the studies of RAPD DNA fingerprinting exhibits that random primers can serve as a very useful tool in distinguishing the strains within species and among genera of edible mushrooms.

The species differentiation in morels by using arbitrary primer based RAPD has been reported by Singh *et al.* (2004). Eight putative species, namely *Morchella esculenta, M. crassipes, M. angusticeps, M. conica, Mitophora semilibera, Morchella spongicola, M. vulgaris* and *Verpa conica* were designated to the sequenced cultures and were analyzed based on similarity with the aligned sequence of BLAST search. They recorded interspecific polymorphism in the

Internal Transcribed Spacer (ITS) regions of 5.8s ribosomal DNA gene of 46 monospore cultures representing species of family Morchellaceae, but no intraspecific ITS polymorphism could be visualized from amongst single and/or different ascospores from various geographical regions. RAPD profiles exhibited significant interspecific polymorphism, distinguishing all the 8 putative species of *Morchella* from each species by primer OPP 06 and a custom primer.

Lakhanpal and Shad (1999) discussed that in spite of all overlaps and intergrades, not only the six species (*Morchella angusticeps, M. esculenta, M. crassipes, M. conica, M. deliciosa* and *M. semilibera*), but also three more seem to be distinct (*M. tibelica, M. rotunda* and *Morchella simlensis* sp. nov.) so long as they are described from mature fruiting bodies. They suggest that the potential of the gene exchange may be used as a means of discriminating taxa within the genus *Morchella.*

Taking the clue from morphological features, the present studies were undertaken to analyse genetically the population of *Morchella* species from Himachal Himalaya.

The details of standard methodology for isolation of genomic DNA, RAPD primers, electrophoresis and statistical analysis of data are given in Annexure-II.

OBSERVATIONS

Quality and Quantity of Genomic DNA

The CTAB procedure used for DNA extraction was able to yield good quality genomic DNA from the fruiting bodies of *Morchella* species. Absolute amount of genomic DNA obtained from fruiting bodies of *Morchella* varied from 14.6–61.3 µg. A_{260}/A_{280} values ranged from 1.8–2.0 indicating the absence of contaminants like phenolics and/or proteins in the DNA preparation. Higher ratios in some samples indicated the presence of RNA. So, RNAase treatment was given to remove the RNA content of the sample.

Random Amplified Polymorphic DNA (RAPD) Analysis

In the present investigations twenty-eight primers were used to amplify the genomic DNA of seven species of *Morchella* and all the primers were able to amplify some regions of the *Morchella* genome. The total numbers of fragments amplified, monomorphic fragments, product size range, minimum and maximum frequency of amplified bands are presented in Table 4.2.

- The 28 primers amplified a total 472 distinct fragments in all the seven species of *Morchella* viz. *M. angusticeps, M. conica, M. crassipes, M. deliciosa, M. esculenta, M. tibelica* and *Morchella simlensis* sp. nov.
- Each primer generated a unique set of amplification products. The size of the amplified products ranged from 105 bp in OPD 03 to 3,628 bp in OPA 06.
- The number of bands for each primer varied from 8 in OPD 15 and OPE 04 to 27 in OPD 11.
- The minimum frequency range of the amplified bands in all the 28 primers was observed to be 14.3%.

Table 4.2 Number of RAPD fragments, size range and resolving power of the random primers used

S.No.	*Primer name*	*Total number of fragments amplified*	*Number of monomorphic fragment*	*Size range in base pairs*	*Minimum frequency of amplified bands*	*Maximum frequency of amplified bands*
1	OPA 06	16	0	575-**3628**	14.3	42.9
2	OPA 08	15	0	570-2729	14.3	57.1
3	OPA 11	17	0	369-2272	14.3	71.4
4	OPA 13	17	0	368-2271	14.3	42.9
5	OPA 14	12	0	310-2306	14.3	42.9
6	OPA 16	10	0	650-2229	14.3	42.9
7	OPA 20	18	0	239-2380	14.3	42.9
8	OPB 01	21	0	248-2727	14.3	85.7
9	OPB 09	18	0	416-2622	14.3	57.1
10	OPB 10	19	0	378-1889	14.3	57.1
11	OPB 12	14	0	377-1624	14.3	57.1
12	OPB 15	14	0	200-1500	14.3	71.4
13	OPB 16	10	0	539-1740	14.3	42.9
14	OPB 19	12	0	409-2170	14.3	42.9
15	OPC 02	13	0	277-1220	14.3	71.4
16	OPC 05	16	0	369-2818	14.3	57.1
17	**OPC 06**	24	**1**	292-2406	14.3	**100**
18	OPC 08	22	0	226-2076	14.3	71.4
19	OPC 15	15	0	321-2358	14.3	42.9
20	OPD 03	22	0	**105**-3135	14.3	71.4
21	OPD 07	21	0	375-1549	14.3	71.4
22	OPD 11	**27**	0	366-2728	14.3	57.1
23	OPD 12	20	0	352-3217	14.3	42.9
24	OPD 15	**8**	0	341-2329	14.3	42.9
25	OPE 01	22	0	355-3154	14.3	42.9
26	OPE 02	21	0	198-2345	14.3	57.1
27	OPE 03	20	0	339-2633	14.3	42.9
28	OPE 04	**8**	0	552-1337	14.3	42.9
	TOTAL	**472**	**1**		**14.3**	**100**

- The maximum frequency range of amplified bands in all 28 primers tested was 100% in primer OPC 06.
- From the total amplified fragments, only one band is monomorphic and rests of all are polymorphic. The band number 15 of the primer OPC 06 having molecular weight 706 bp was monomorphic. This monomorphic band inferred that this locus is common to all seven species of *Morchella*. This shows that this locus may belong to some indispensable gene. If this locus can be further cloned and sequenced, then this locus may be further characterized. (Table 4.3)

Table 4.3 Names and sequences of the random primers used in this study

S. No.	*Primer name*	*Primer sequence (5'-3')*
1	OPA 06	GGTCCCTGAC
2	OPA 08	GTGACGTAGG
3	OPA 11	CAATCGCCGT
4	OPA 13	CAGCACCCAC
5	OPA 14	TCTGTGCTGG
6	OPA 16	AGCCAGCGAA
7	OPA 20	GTTGCGATCC
8	OPB 01	GTTTCGCTCC
9	OPB 09	TGGGGGACTC
10	OPB 10	CTGCTGGGAC
11	OPB 12	CCTTGACGCA
12	OPB 15	GGAGGGTGTT
13	OPB 16	TTTGCCCGGA
14	OPB 19	ACCCCCGAAG
15	OPC 02	GTGAGGCGTC
16	OPC 05	GATGACCGCC
17	OPC 06	GAACGGACTC
18	OPC 08	TGGACCGTCC
19	OPC 15	GACGGATCAG
20	OPD 03	GTCGCCGTCA
21	OPD 07	TTGGCACGGG
22	OPD 11	AGCGCCATTG
23	OPD 12	CACCGTATCC
24	OPD 15	CATCCGTGCT
25	OPE 01	GTTTCGCTCC
26	OPE 02	GGTGCGGGAA
27	OPE 03	CCAGATGCAC
28	OPE 04	GTGACATGCC

a) RAPD profiles of Morchella species with OPA series of primers

Seven decamer primers of OPA series were used, viz. OPA 06, OPA 08, OPA 11, OPA 13, OPA 14, OPA 16, and OPA 20. Altogether they produced 104 amplified fragments and all of them were polymorphic (Plate–I a).

b) RAPD profiles of Morchella species with OPB series of primers

Seven decamer primers of OPB series were used, viz. OPB 01, OPB 09, OPB 10, OPB 12, OPB 15, OPB 16, and OPB 19. They produced 108 amplified fragments altogether and all of them were polymorphic.

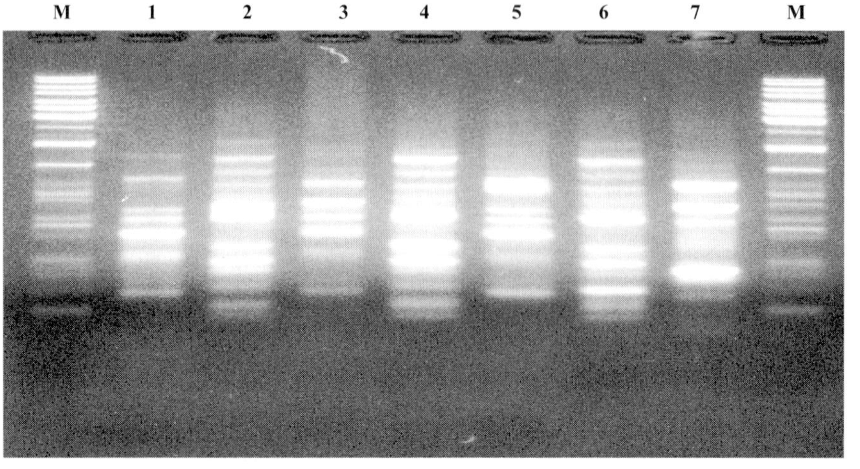

(a)

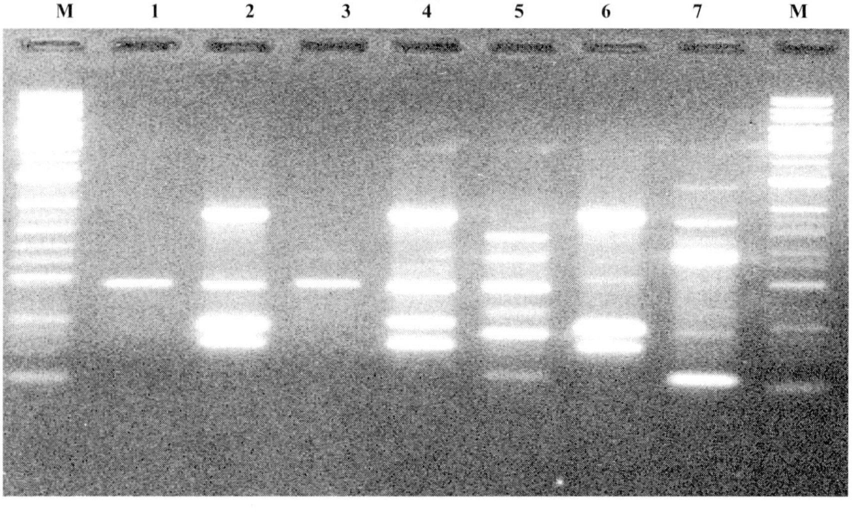

(b)

PLATE–I (a, b) RAPD Profiles of *Morchella* spp. generated by Random primers (OPA 13 and OPC 02) **M** represents 100 bp DNA Ladder (MBI, FERMENTAS, Inc. 7520 Connelley Drive, Suite A, Hanover, MD, 21076, USA) Lane **1–7** represents following *Morchella* species:

1. *M. angusticeps;* 2. *M. esculenta;* 3. *M. tibelica;* 4. *M. deliciosa;* 5. *M. conica;* 6. *Morchella simlensis sp. nov.;* 7. *M. crassipes*

c) RAPD profiles of Morchella species with OPC series of primers

Five decamer primers of OPC series were used, viz. OPC 02, OPC 05, OPC 06, OPC 08 and OPC 15. They produced 90 amplified fragments altogether, out of which only one fragment was monomorphic (OPC 06, Band number - 15), and rest all were polymorphic (Plate–I b).

d) RAPD profiles of *Morchella* species with OPD series of primers

Five decamer primers of OPD series were used, viz. OPD 03, OPD 07, OPD 11, OPD 12 and OPD 15. They produced 98 amplified fragments altogether and all were polymorphic. Minimum and maximum number of RAPD fragments were obtained in this experiment from OPD 15 (8) and OPD 11 (27).

e) RAPD profiles of *Morchella* species with OPE series of primers

Only four decamer primers of OPE series were used, viz. OPE 01, OPE 02, OPE 03, and OPE 04. Together they produced 71 amplified fragments and all were polymorphic. Minimum numbers of RAPD fragments were also produced in this experiment by OPE 04 (8).

SIMILARITY ANALYSIS AND CLUSTERING OF DATA

The phylogenetic tree (dendrogram) generated was based on the data resulting from RAPD analysis (Fig. 4.2). At 64% similarity coefficient, all the seven species were separated in seven different clusters (Fig. 4.1). The resulting dendrogram clustered *Morchella* species into two major groups: A and B at 15% similarity level which are further subdivided into different subgroups.

(i) Group A

Group A included three species of morels, viz. *M. angusticeps, M. conica* and *M. tibelica* clustered together as black morels. All these three black species grouped at 24% similarity

	C1	C2	C3	C4	C5	C6	C7
C1	1.00						
C2	0.12	1.00					
C3	0.30	0.18	1.00				
C4	0.16	0.59	0.1909	1.0000			
C5	0.36	0.11	0.1918	0.1790	1.0000		
C6	0.1308	0.5698	0.1756	0.6434	0.1511	1.00 00	
C7	0.0921	0.2539	0.1494	0.2898	0.1712	0.27 35	1.0000

Fig. 4.1 Similarity coefficient between different *Morchella* species
C1 – *Morchella angusticeps,* **C2** – *Morchella esculenta,* **C3** – *Morchella tibelica,* **C4** – *Morchella deliciosa,* **C5** – *Morchella conica,* **C6** – *Morchella simlensis* sp. nov., **C7** – *Morchella crassipes*

level into two subgroups. Dendrogram showed that *M. angusticeps* and *M. conica* have 36% similarity level while *M. tibelica* had 24% similarity level with this group. This concludes that *M. angusticeps* and *M. conica* were closely related than *M. tibelica.*

(ii) Group B

This group includes four species of *Morchella,* viz. *M. esculenta, M. deliciosa, M. crassipes* and

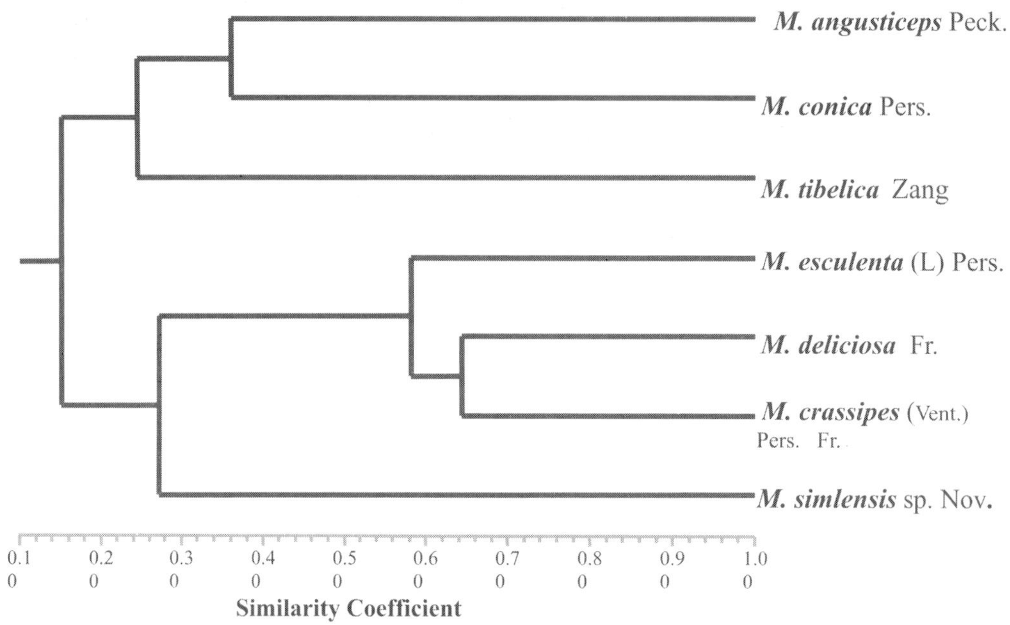

Fig. 4.2 Dendrogram of *Morchella* species showing clustering of RAPD patterns

Morchella simlensis sp. nov. All the four species clustered at 28% similarity level into two sub-groups. *M. deliciosa* and *M. crassipes* were closely related (64% similarity level) out of all the four species. *M. esculenta* had 58% similarity level with this group. All the three species viz. *M. deliciosa, M. esculenta* and *M. crassipes* were only 28% similar to newly categorized species, *Morchella simlensis* sp. nov. This newly discovered species (with white ridges, Shad, 1989) seems to be similar to the group B – yellow morels on the basis of genotypic diversity.

Therefore, the phylogenetic tree presented here suggested that the group A (black morels – *M. angusticeps* and *M. conica*) and group B (yellow morels – *M. esculenta, M. deliciosa* and *M. crassipes*) have been reported to be separate taxonomic groups and were distantly related. The other two species: *M. tibelica* and *Morchella simlensis* sp. nov. have also been categorized to these respective groups. *Morchella simlensis* sp. nov. has been more conspecific to yellow morels (28% similarity level) and *M. tibelica* showed similarity to black morels (24% similarity level). Thus, the conclusion was drawn from the phylogenetic tree that the black species (*M. angusticeps, M. conica* and *M. tibelica*) demonstrated higher interspecific variability than yellow morels (*M. esculenta, M. deliciosa* and *M. crassipes*).

DISCUSSION

There is very little consensus on the systematic status of the species composing the genus *Morchella,* because the traditional methods of classification of *Morchella* genus, based on ascocarp morphology, have become ambiguous and now more reliable methods of phylogenetic analysis are needed. The ascocarps of *Morchella* are highly polymorphic in appearance, exhibiting many shapes, sizes and colour variations. This polymorphic nature of the genus has contributed to taxonomic uncertainties in *Morchella* (Bunyard *et al.*, 1994). Many researchers conducted many polygenetic studies on the available species of *Morchella* to detect the genetic variability/diversity.

In the present investigations the polymorphism is assessed through Random Amplified Polymorphic DNA (RAPD) markers. The phylogenetic tree presented here (Fig. 4.2) suggested that *Morchella* species fall into two groups – black morels and yellow morels. Therefore, black morels (*M. angusticeps, M. conica* and *M. tibelica*) and yellow morels (*M. esculenta, M. deliciosa, M. crassipes* and *Morchella simlensis*) were separate taxonomic groups.

A number of investigations in the recent past (Gessner *et al.*, 1987; Royse and May, 1990; Bunyard *et al.*, 1994, 1995; and Wipf *et al.*, 1996, etc.) have attempted to resolve the taxonomic problems in the genus by the use of various molecular techniques, but there has been scarcity of data on genotypic diversity of morels in India (Sharma *et al.*, 2003, 2004).

The present investigations supported the findings of some previous researchers (Gessner *et al.*, 1987; Royse and May, 1990; Yoon *et al.*, 1990; Bunyard *et al.*, 1994, 1995). Gessner *et al.* (1987) investigated the allelic variation and segregation in *M. esculenta* and *M. deliciosa* and reported them as Mendelian populations, which can be reproductively isolated. Royse and May (1990) used allozyme variation to show the polymorphism. They showed evidence in support of a few polymorphic species within the genus as determined by multi-locus enzyme electrophoresis as black and yellow morels.

The systematic assessment of *Morchella* species using RFLP of 28S ribosomal RNA gene was reported by Bunyard *et al.* (1994). The phylogenetic tree developed from the restriction data revealed that the black morels (*M. angusticeps, M. elata* and *M. conica*) and the yellow morels (*M. esculenta, M. crassipes* and *M. deliciosa*) were separate taxonomic groups. Later on in 1995, they reported the inter- and intraspecific polymorphism among different putative species of *Morchella* (including *M. semilibera*) and elucidated the relationships of *Morchella* to other genera of Morchellaceae on the basis of restriction analysis of the 28S Ribosomal RNA gene. The phylogenetic tree (dendrogram) drawn suggested that black morels (*M. angusticeps, M. elata* and *M. conica*), yellow morels (*M. esculenta, M. crassipes* and *M. deliciosa*) and semi-free capped morels (*M. semilibera*) were separate, polymorphic and taxonomic groups. The black morels grouped together had much more variability and yellow species showed high degree of conspecificity. Therefore, black morels demonstrated as much variation interspecifically, as intraspecifically. In the present investigations, the phylogenetic tree drawn through RAPD suggested that the black and yellow morels were separate taxonomic groups. The observations made in the present studies were concordant to that of Bunyard *et al.* (1994 and 1995) inferring that black morels grouped together had much more variability than yellow morels.

Buscot *et al.* (1996) studied the DNA polymorphism within morels species by PCR/RFLP analysis of the ribosomal DNA spacers and microsatellites-primed PCR. They recognized the two morphological groups of *Morchella* species—(i) adnate morels (base of the hymenium is not free) to which *Morchella esculenta* and *M. crassipes* belong and (ii) the distant species (base of hymenium is free) to which *M. conica* and *M. angusticeps* belong. Again in 1996, Wipf *et al.* reported DNA polymorphism in morels by complete sequences of the Internal Transcribed Spacer (ITS) of genes coding for rRNA in *Morchella esculenta* (yellow morels) and *M. conica* (black morels). They inferred that both the groups are distinct ones. The genetic diversity within and between populations of *M. esculenta* by using RAPD markers was observed by King and Jacobson (2000). They tested the null hypothesis that *M. esculenta* populations in Iowa would exhibit low levels of genetic diversity using RAPD markers and inferred that morel population might be highly inbred.

Recently, the species differentiation in morels *Morchella esculenta, M. crassipes, M. angusticeps, M. conica, Mitophora semilibera, Morchella spongicola, M. vulgaris* and *Verpa conica* using arbitrary primer based RAPD has been reported by Singh *et al.* (2004). These eight putative species were designated to the sequenced cultures and were analyzed on the basis of similarity with the aligned sequence of BLAST search. They recorded interspecific polymorphism in 46 monospore cultures representing species of family Morchellaceae, but no intraspecific ITS polymorphism could be visualized among single and/or different species from various geographical regions.

The phylogenetic tree drawn in the present studies (Fig. 4.2) shows a high degree of conspecificity between yellow morels viz. *M. esculenta, M. crassipes, M. deliciosa,* and *Morchella simmlensis,* respectively. This confirms that yellow species have been intraspecifically very less variable. The black species of morels (*M. angusticeps, M. conica,* and *M. tibelica*) have been grouped together, but they have much more variability than yellow morels. Thus, black group of morels demonstrated higher intraspecifically variability corroborating observations of Bunyard *et al.* (1994, 1995), Wipf *et al.* (1996), King and Jacobson (2000) and Singh *et al.* (2004). They observed high degree of conspecificity of *M. esculenta, M. crassipes* and *M. deliciosa* (yellow morels). Even the isolates of these species were grouped together with little variations, most being indistinguishable from each other. In addition, much variation was seen interspecifically as intraspecifically. But black species (*M. angusticeps, M. elata* and *M. conica*), grouped together, has much more variability than yellow species.

The previous analysis of genotypic diversity of morels have been reported by various techniques earlier by isozyme analysis (Kulkarni and Kamerath, 1989; Yoon *et al.*, 1990 and Wipf *et al.*, 1996 etc.); allozyme variations (Royse and May, 1990); ELISA (Jung *et al.*, 1993); RFLP of ribosomal DNA spacers (Buscot *et al.* 1996); ITS gene coding for rRNA (Wipf *et al.*, 1996) etc. mainly, but in the present investigations, RAPD has been followed. This technique was also used earlier in determining the genotypic diversity in *Morchella* species (King and Jacobson, 2000 and Singh *et al.*, 2003, 2004). RAPD techniques have several advantages over established RFLP methodologies. RAPD has simple experimental setup, requires small amount of genomic DNA, results are quick, efficient screening for DNA based polymorphism, and no radioisotopes are required. These advantages, together with the relative ease of the technical procedure, led to the increased use of RAPD in Genetic Divergence Analysis.

Thus, on the basis of RAPD markers, an attempt has been made to resolve the taxonomic problems or polymorphisms in different members of family Morchellaceae and broadly they can be categorized into two groups—black and yellow morels.

CHAPTER

5 Ecological Aspects

CHAPTER OUTLINE

- Introduction and Review of Literature
- Observations and Discussion

INTRODUCTION AND REVIEW OF LITERATURE

The Himalayan forests harbour diverse forms of mushrooms. The mushroom mycoflora shows changes with different forest types, altitude, soil, and other related factors. Different tree species serve as indicators for different species of mushrooms. A difference is apparent in species which seem to prefer raw humus, such as is found in conifer forests where the needles decompose slowly, and those which prefer mild humus, such as that found in deciduous forests.

Among the various types of wild mushrooms, morels (*Morchella* spp.) occupy a special position in the forests of Himachal Pradesh. Himachal Pradesh ranks first in the production of morels in India. It is the excellent flavour that the morels possess, which has fascinated man and made him interested in collecting morels for food (Rofle and Rofle, 1925; Lambert, 1983 and Gray, 1959). But despite the long standing interest in morels, the method for morel cultivation still remains undefined, and the critical condition, which favours the carpophore development, is yet to be fully determined in their life cycle. This is the reason why only sporadic success has been achieved in their artificial cultivation (Cailleux, 1969; Delmas, 1978; Ower, 1982). Gray (1959) felt that the problem is exceptionally challenging one in the field of applied mycology and discovery of a successful method of morel cultivation will give a big fillip to mushroom industry.

Among the physical factors in the natural environment affecting the larger fungi, the important fungi, the important ones are moisture, composting of substrates, pH, available nitrogen and mineral status (Grainger, 1946). Munjal *et al.* (1977) and Kaul *et al.* (1981) also emphasized the significance of some of these factors in the appearance of morels in their studies in Himachal Pradesh and Kashmir valley, respectively.

Ecological data on microfungi is available in plenty, but ecology of macrofungi in general has received little attention from mycologists (Graham, 1927; Willkins *et al.*, 1937, 1938, 1939, 1940 and 1946; Corner, 1934; Grainger, 1946; Hawker, 1954, 1957 and Hora, 1959, 1972). This is because of the ephemeral nature and uncertainty of their appearance and incomplete knowledge of fungal systematics and difficulties in the identification of larger fungi. Shantz and Piemeisel (1917) published probably the first paper on the ecology of mushrooms, wherein they discussed fungus fairy rings in Eastern Colarado and their effect on vegetation.

Graham (1927) discussed the various aspects of fungus ecology in the Chicago region. Later, Wilkins *et al.* (1937, 1938), Wilkins and Patrick (1939, 1940) and Wilkins and Harris (1946) published a series of papers on the ecology of larger fungi. Slipp and Snell (1944) treated the Boletaceae of the Kaniksu National Forest in terms of species distribution within different successional phases of the Thiya–Tsuga zone. They emphasized the mycorrhizal relations of the members of the Boletaceae with different trees. Kumar *et al.* (1990) conducted ecological studies in some Agaric families, recording data on fruiting period and their association with different tree species in different forest types.

With regard to morels, McCubbin (1913); Moser (1949 a, b); Ramsbottom (1953) and Baker and Matkin (1959) reported excessive growth of *Morchella* on disturbed ground, bombed sites, burnt places and where burnt ash of paper or wood had been added. Baker and Cook (1974) investigated that fruitings of morels in spring may be due to their ability to grow at low temperatures to the exclusion of competition. Delmas and Bunel (1974) pointed out (after Heim, 1936, 1941; Cailleux, 1956, 1969) that one of the difficulties in cultivating morels lies in the ecological specificity of each species of morels. They provided a list of basic facts on ecology of morels, e.g. types of soils, associated trees, altitude, season of appearance, etc. and emphasized that their trials on various reconstituted natural media had yielded no fruiting bodies (see Chang and Hayes, 1978).

Schmidt (1983) studied the life cycle and ecology of morels in the field. He observed that the ascospores of *Morchella esculenta* germinated at temperatures as low as 20°C *in vitro,* and the spores buried at morel sites in cellophane bags swelled but did not produce germ tubes until the soil temperatures exceeded 10°C, approximately. But the spore germination was inhibited as the soil continues to warm after fruiting body emergence, above 10°C and the spore did not remain viable after one year near the soil surface. He reported that sterile rye buried in the field was readily colonized by *Morchella esculenta* when cased with site soil containing either spores or sclerotia at soil temperatures below 10°C, but as temperature increased, only other saprophytes appeared.

Kaul (1975) analyzed forty soil samples from morel bearing sites and burnt situations for major inorganic components, organic matter and moisture in order to determine the factors responsible for carpophore formation. The results indicated that higher calcium contents of soil and alkaline reaction of the substrate may have a role in carpophore development. Munjal *et al.* (1977) studied the edaphic factors in relation to growth of *Morchella* species. They observed that *M. deliciosa* is adapted to low elevation zone and prefers loamy soil, while *M. angusticeps* and *M. conica* grow at higher elevations, indicating that they require cooler climatic conditions and more aeration. The analysis of the soil samples of different substrates showed that *Morchella* species prefer higher pH (8.1). This supports the view of Repin (1901), who claimed to get more fruit bodies of *Morchella* around a bed composed of dry leaves of apple deposited in a trench and rendered alkaline with sodium carbonate.

Kaul *et al.* (1981) investigated the morel bearing sites in Kashmir valley and noted that morel fruit bodies were found when soil moisture was above 20%, soil temperature ranged between 10-23°C above ground and 8.5–20°C below ground, relative humidity 58 to 77%, pH neutral or alkaline and when there was high level of available nitrogen. They also observed higher levels of calcium, potassium and phosphorus in soil during the fruiting season.

The details of standard methodology for soil analysis, estimation of nitrogen, phosphorus, calcium, magnesium and potassium are given in Annexure–III.

Considering the available data on ecology of morels, the ecological aspects of morels were investigated under a holistic package of following parameters:

i. Characteristics of morel bearing and non-morel bearing soils.
ii. Association of morels with different forest types.
iii. Distribution of morels species in different altitudinal/vegetational zones.
iv. Quantification of the duration of appearance of different species of morel.
v. Soil microflora of morel bearing and non-morel bearing soils.
vi. pH of soil samples.

OBSERVATIONS AND DISCUSSION

i. Characteristics of Morel Bearing and Non-Morel Bearing Soils

The soil samples have been collected and analyzed from different types of forests starting with chir pine to fir-spruce, ranging in altitude from 1000 m – 2500 m.

a) Non-morel bearing soils (Table 5.1)

The colour of the soil in chir forest was nought to dark blonde ($5D_3$ – $5D_4$). The average pH of the soil was 6.9. The soil was loamy and comparatively low in organic matter. The average nitrogen, phosphorus, potassium, sodium, calcium and magnesium contents were 42.7, 10.0, 35.12, 5.76, 80.0 and 46.3 mg/100 gm of the soil, respectively.

Table 5.1 Composition of non-morel bearing soils

Vegetation zone	*pH*	*N mg/ 100 gm*	*P mg/ 100 gm*	*Ca mg/ 100 gm*	*K mg/ 100 gm*	*Mg mg/ 100 gm*	*Na mg/ 100 gm*
Chir forest	6.8	42.70	17.00	80	35.12	46.30	5.76
Deodar forest	7.0	59.56	27.68	130	44.87	44.87	6.51
Deodar-Spruce forest	6.7	56.83	29.76	115	42.63	60.00	7.65
Abies-Spruce forest	6.9	57.62	41.43	100	51.23	32.00	8.00
Oak forest	6.5	69.0	39.53	90	55.56	30.00	8.25

In the deodar forest, the soil was chocolate brown to dark brown ($5F_3$ – $5F_4$), sandy loam and comparatively rich in organic matter compared to that of chir forest soil. The nitrogen, phosphorus, potassium, calcium, magnesium and sodium contents were 59.56, 27.68, 44.87, 130.00, 42.00, and 6.57 mg/100 gm of the soil, respectively. The average pH of the soil was 7.0.

In deodar-spruce forest, the soil was chocolate brown in colour. It was sandy loam and rich in organic matter. The average pH was 6.7. Average nitrogen, phosphorus, potassium, sodium, calcium and magnesium contents were 56.83, 29.76, 42.63, 7.65, 115.00, and 60.00 mg/100 gm of the soil, respectively.

In silver fir-spruce forest, the colour of the soil was chocolate brown. It was sandy loam and rich in organic matter. The average pH was 6.9. Average nitrogen, potassium, phosphorus, magnesium, calcium and sodium contents of the soil were 57.62, 51.23, 41.43, 32.00, 100.00 and 8.00 gm/100 gm of the soil, respectively.

In oak forest, the colour of the soil was brown. The average pH of the soil was 6.5. Average contents of nitrogen, potassium, phosphorus, magnesium, calcium and sodium were 69.0, 55.56, 39.53, 30.0, 90.0 and 8.25 mg/100 gm of the soil, respectively.

b) Morel bearing soils (Table 5.2)

In the morel bearing soil of the chir forest, the average pH of the soil was 7.2. The average nitrogen, phosphorus, potassium, sodium, calcium contents were 44.7, 17.5, 6.75, 100 and 46.0 gm/100 gm of the soil, respectively.

Table 5.2 Composition of morel bearing soils

Vegetation zone	*pH 100 gm*	*N mg/ 100 gm*	*P mg/ 100 gm*	*Ca mg/ 100 gm*	*K mg/ 100 gm*	*Mg mg/ 100 gm*	*Na mg/*
Chir forest (*P. roxburghii*)	7.2	44.70	17.50	100	35.00	46.00	6.75
Deodar forest (*Cedrus deodara*	8.5	62.50	28.00	160	43.20	44.25	7.50
Deodar-Spruce forest (*Cedrus deodara* and *Picea smithiana*)	8.2	60.25	30.25	140	48.50	65.00	6.25
Abies-Spruce forest (*Picea smithiana*)	8.0	58.62	44.00	130	50.00	40.00	8.40
Oak forest 6.8 (*Quercus* spp.)	72.70	42.00	105	58.40	35.00	8.10	

In deodar forest, the average pH of the soil was 8.5. The average nitrogen, phosphorus, potassium, sodium, calcium and magnesium contents were 62.50, 28.0, 43.20, 7.50, 160.0 and 44.25 gm/100 gm of the soil, respectively.

The average pH of such soils in deodar-spruce forest was 8.2. The average contents of nitrogen, phosphorus, calcium, potassium, magnesium and sodium were 60.25, 30.25, 140.0, 48.50, 65.0 and 8.25 mg/100 gm of the soil, respectively.

The average pH of morel bearing soils in silver fir–spruce forest was 9.0. Average nitrogen, phosphorus, calcium, potassium, magnesium and sodium content were 58.62, 44.0, 130.0, 50.0, 46.0 and 8.4 mg/100 gm of the soil, respectively.

The average pH of morel bearing soils of oak forest was 6.8. Average nitrogen, phosphorus, calcium, potassium, magnesium and sodium content were 72.70, 42.0, 105.0, 58.40, 35.0 and 8.1 mg/100 gm of the soil, respectively.

Comparison of pH values of morel bearing and non-morel bearing soils in the different localities revealed that morel bearing soils have higher pH values ranging from slightly acidic to alkaline (6.8 – 8.5). This supports the view of Repin (1901), who claimed to get fruit bodies of *Morchella* around a bed composed of dry leaves of apple deposited in trench, rendered alkaline with sodium carbonate.

Munjal *et al.* (1977) also reported that *Morchella* species prefer a higher pH (8.1). Kaul (1975) observed abundant development of morels on burnt sites in forest probably because burnt soils have higher pH and calcium content. He also reported that spore germination in *Morchella* species is favoured by alkaline medium and that all the species of *Morchella* require alkaline conditions.

There was a gradual increase in the average levels of nitrogen, phosphorus, calcium, potassium, magnesium and sodium in the soil from chir to oak forest. The calcium, nitrogen, phosphorus, potassium, magnesium and sodium content were higher in the morel bearing soils than the non-morel bearing soils of different forest types. The calcium content was comparatively higher in deodar forest, and incidentally deodar forest also showed highest production of *Morchella*. Kaul *et al.* (1981) observed high levels of available nitrogen, phosphorus and potassium in soil samples analysed from morel bearing sites in Kashmir valley. It is difficult to draw any concrete conclusion, but it is evident that favourable pH and calcium content of the soil favour better morel production.

ii. Morel Distribution in Different Forest Types and at Different Elevation

a) Habitat of Morchella spp. (Plate I a-e, II a-c)

Morels commonly occur in soils rich in humus and humidity. These are found in pure as well as mixed coniferous forests of *Pinus roxburghii, P. wallichiana, Cedrus deodara, Picea smithiana* and *Abies pindrow*. They also occur in *Quercus* forests, grasslands, pastures and apple orchards and occasionally in wheat fields. They are usually terricolous and lignicolous. By carefully digging the fruit bodies in the field it has been observed that most of the *Morchella* species enter into mycorrhizal association with ferns, strawberry, raspberry plants and grass roots (See Chapter 12). The mycelial cord in some of the species has been traced into the mycorrhizal roots by careful removal of soil bit by bit. In *M. semilibera* no such relationship has, however, been observed. It has been collected only from *Abies–Picea* forest mixed with chestnut, especially under decomposed leaves of the latter.

b) Fruiting period of Morchella

Observations recorded in Fig. 5.1 and Table 5.3 reveal that at lower elevations *Morchella* species started appearing around mid-February and were collected upto the end of April. At mid-elevations, the collections started in early March and extended upto mid-May. At higher elevations, they appear during May-June. These areas are usually longer snowbound and morels start appearing only when snow begins to melt. They occur first along the line of melting snow under forest debris, such as broken twigs and small branches, etc. (Fig. 5.2).

The maximum morel collections are usually made during spring, i.e. March and April. During these months a number of people wander in search of morels, and at times the competition among collector is tough as the area of collector is limited.

Since morels appear only after snow starts melting and when there is at least some spring rain, it lends support to the speculation of Schisler and Baker (c.f. Baker and Cook, 1974) that the fruiting of morels in spring may be due to their ability to grow at low temperatures to exclude competitions. Delmas and Bunel (1974) have also recorded January to June, as the fruiting season for *Morchella* spp. in France.

c) Distribution of Morchella species in different months and in relation to altitude and forest types (Fig. 5.1 and Table 5.4)

Among the species of morels, *M. deliciosa* occurs at lower elevations, while *M. angusticeps, M. tibelica, M. semilibera* and *M. conica* at higher elevation. *M. crassipes* and *Morchella simlensis* were collected from 2000–2200 m, whereas *M. esculenta* was found at all elevations, ranging from 1650 – 2800 m. Maximum collections were made in deodar forests, followed by deodar-spruce, chir pine and oak. March and April were the most productive months for morel collections, followed by May, June, February and July (Fig. 5.3).

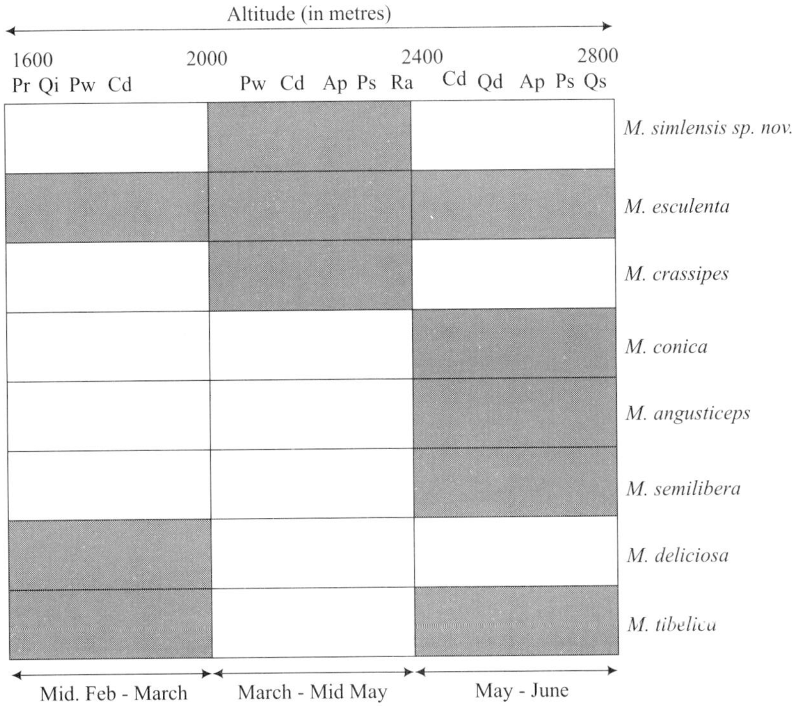

Pr - *Pinus roxburghii*
Qi - *Quercus incana*
Pw - *Pinus wallichina*
Cd - *Cedrus deoxarda*
Ap - *Abies pindrans*
Ps - *Picea smithiana*
Ra - *Rhododenderon arborcum*
Qd - *Quercus dilatata*
Qs - *Quercus semicarpitolia*

Fig. 5.1 Distribution of *Morchella* species in relation to altitude and fruiting season

d) Colour variations in species of Morchella: (Table 5.5, Plate II d, e)

The colour of the pileus of fruit bodies in *Morchella* species changed with forest type and soil. The pileus colour of *M. esculenta* was yellowish to whitish in chir forests, yellowish to brownish in deodar forest, brown in deodar–spruce forest, dark brown in silver fir–spruce forest and brownish to yellowish in oak forest. The pileus of other species also varied similarly with vegetation type. What actually determines or controls the colour is not known, but because the colour changes with the forest type and it has been observed that soil type also varies with the forest type, the colour of fruiting bodies may possibly depend both upon forest and the soil type.

Among different species collected, *M. deliciosa* was recorded at lower elevations (1600 – 2000 m), supporting the observations of Munjal *et al.* (1977). *M. angusticeps, M. tibelica, M. semilibera,* and *M. conica* occupy niches at higher elevations (Munjal *et al.*, 1977). *M. crassipes*

Table 5.3 Fruiting period of *Morchella* species in nature

Locality	*Altitude forest type*	*Fruiting period*
Bahu, Kandugad	1600 m Mixed	Mid-Feb. to March
Kungush, Tara Devi, Hatkoti	2000 m Coniferous (Chir-Oak)	
Summer Hill, Chowi, Chajpur, Nankheri	2000 m Coniferous (Pine)	March to mid-May
Manali, Sarhan, Chopal, Sungra, Baghi, Khadrala	2400 m Spruce (Deodar)	
Jaloripss, Nahuni, Hatoo, Churdhar	2400 m Mixed (Abies-brown - 2800 m Oak, and *Rhododenderon*)	May to June

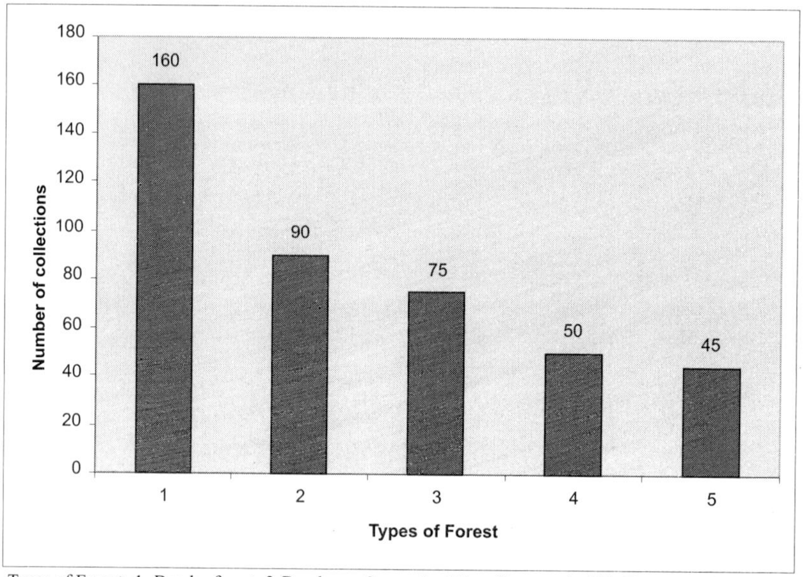

Types of Forest: 1- Deodar forest, 2-Deodarom Spruçe 3- Abies – Spruce , 4 - Chir Forest, 5 - Oak Forest

Fig. 5.2 Number of collections of *Morchella* made in different forests

and *Morchella simlensis* were collected from 2000 - 2200 m, whereas *M. esculenta* was found at all elevations, ranging from 1650–2800 m. *M. esculenta,* therefore, appears to be a more versatile species. Similar observations have been made by Delmas and Bunel (1974), who recorded *M. deliciosa* at an altitude of 800–1000 m, *M. crassipes* at 1200–1500 m, *M. elata* at 400–1500 m, *M. costata* upto 600 m, and *M. umbarina* upto 1000 m. *M. vulgaris, M. rotunda, M. spongicola* and *M. hortensis* were recorded in the plains.

These studies clearly point out the relationship between species distribution and the altitude, and such relationship has been repeatedly observed during the past three decades.

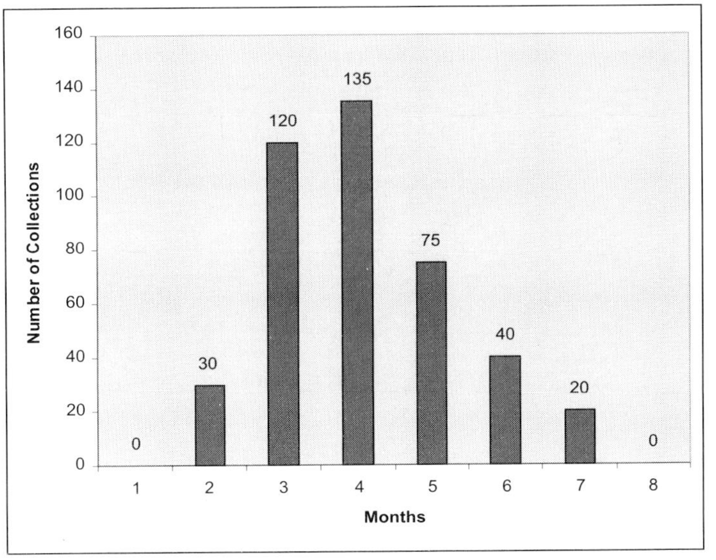

Months: 1 – January, 2 – February, 3 – March, 4 – April, 5 – May, 6 – June, 7 – July, 8 – August

Fig. 5.3 Monthly collections of *Morchella*

Table 5.4 Distribution of *Morchella* species in relation to altitude

Localities	*Altitude in m*	*Morel species*
Manikaran, Bahu, Odi	1600	M. *deliciosa*
Kandugad, Tara Devi, Hatkoti	2000	*M. esculenta*
Khanag, Chowi, Summer Hill, Barubag	2000	*M. esculenta, M. simlensis*
Sungra, sangla, Chopal, Sarhan, Nankheri	2400	*M. crassipes*
Luhal, Narkanda, Kutwa, Sarahan, Bagi	2400	*M. conica, M. esculenta*
Kharrala, Nichar, Panju	2800	*M. angusticeps, M. tibelica, M. semilibera*
Nanu, Below Hatoo peak, Below Jalori Pass, Jakhu, Churdhar, Panju	2800	*M. esculenta, M. conica, M. angusticeps, M. semilibera, M. tibelica*

As for the figures of collections in different forest types and during different months, *Cedrus deodara* and *Cedrus-Spruce* forests were found to be most productive Table (5.7), whereas among months, March–April yielded the maximum number of collections (Table 5.6). But these are relative figures and cannot be absolute because of the competitions among the collectors. However, these figures are certainly indicative of the preponderance of *Morchella* in these types of forests and during these months, because if these were not so abundant, so many collections would not have been possible by the researchers.

Table 5.5 Variation in colour with soil and forest type in species of *Morchella*

Forest type	*Soil characteristics*	*Colour of species*
Chir forest	Loamy soil comparatively poor in organic matter	Yellowish *M. deliciosa* + whitish *M. esculenta*
Deodar forest	Sandy loamy, soil with low organic matter content	Yellowish *M. deliciosa* to brownish *M. conica* + *M. crassipes* + *M. angusticeps* + *Morchella simlensis*
Deodar-Spruce forest	Sandy loam, rich inorganic matter	Brownish *M. esculenta* + *M. conica* + *M. angusticeps* + *M. semilibera* + *M. tibelica*
Abies-Spruce forest	Sandy loam, rich inorganic matter	Dark brown *M. esculenta* + *M. conica* + *M. angusticeps* + *M. semilibera*
Oak forest	Sandy loam, rich inorganic matter	Brownish to yellowish *M. esculenta* + *M. conica*

Table 5.6 Number of collections of *Morchella* made in different forests

Forest Type	*Number of collections made*
Chir forest	49
Deodar forest	160
Deodar-Spruce forest	95
Abies-Spruce forest	75
Oak forest	41
Total Collections	**420**

Table 5.7 Monthly collections of *Morchella* made

Month	*Number of collections*
February	30
March	120
April	135
May	75
June	41
July	19
Total Collections	**420**

PLATE I (a-g) (*a*) *Morchella esculenta and Morchella angusticeps* in natural grassy habitats in an apple orchard. (*b*) *Morchella tibelica* growing with raspberry plants in an apple orchard. (*c*) *Morchella deliciosa* growing with raspberry plants in an apple orchard. (*d*) *Morchella crassipes* growing with grasses and raspberry plants in an apple orchard. (*e*) Fructification of *Morchella crassipes* forming mycelial strands with grass. (*f*) Fructification of *M. deliciosa* forming mycelial strands with *Oxalis.* (*g*) Uprooted fruiting body of *Morchella conica* showing mycelial strands with the substratum.

PLATE II (a-d) (a) Collection of morels in the natural habitat. Arrow (↓) the niches where the accompanying morel fruit bodies were found. (b) *Morchella* fruiting body growing on wood (lignicolous). (c) *Morchella* fruiting body growing on pine needle litter. (d) Colour variations in the fruit bodies changing with the type and colouration of the substrate(s).

CHAPTER

6 Physiological Aspects of Morels

CHAPTER OUTLINE

(I) GENERAL PHYSIOLOGICAL ASPECTS

INTRODUCTION

The exploitation of domestication technology of morels is well known to be the most challenging problem because the environmental parameters and physiological factors involved in the control of their morphogenesis and production of fruiting bodies are unknown and extremely difficult to elucidate. Cultural studies on morels stem from a desire to obtain fruiting bodies in some way. Therefore, they cover a wide variety of ideas and experiments indoors and outdoors and on different types of substrates, natural as well as semi or synthetic ones, all indicating attempts towards achieving cultivation rather than truly cultural studies.

Brock (1951) for the first time studied the nutritional requirements of *Morchella esculenta* and recorded that the cultures could grow on agar or in liquid culture on synthetic media composed of mineral salts and suitable carbon or nitrogen sources. He observed that *M. esculenta* grew well on starch, maltose, fructose, turanose, glucose and sucrose, and moderately well on several others. Polyhydric alcohols and L-methyl sugars supported little growth. Insulin was utilized in growth on agar but not in liquid culture. He found that the nitrogen sources, cysteine-HCl, aspartic acid, asparagine, urea, sodium nitrite, various ammonium salts, alanine, glutamic acid-HCl, and sodium nitrate were favourable. Other amino acids and nitrogen compounds were intermediate or poor. Ammonium citrate, thiourea, hydroxylamine-HCl, and hydrazine-2 HCl were toxic. He recorded 6.93 as optimum pH for the growth of mycelium of *M. esculenta*.

Robbins and Hervey (1959) studied the growth of *Morchella esculenta* and *M. crassipes* in a medium of mineral salts, dextrose, casein hydrolysate, B-vitamins and purine and pyrimidine bases at pH 4.5. It was markedly increased by the addition of natural preparations, especially wood extract was effective also at neutral reactions. They explained that wood extract and other substances supply *Morchella* with a natural unidentified growth substance (or more than one) for which *Morchella* is partially deficient. They also recorded that out of two species—*M. crassipes* and *M. esculenta*, only *M. crassipes* formed sclerotia on 2% malt extract.

Hervey *et al.* (1978) explained that the ascospores have no dormant period. They probably germinate immediately and grow during the summer months in soil temperature of upto 20°C. They stated that the over wintering stage may form sclerotia, which would then be the source of the phase which produces fruits in the spring. Gilbert (1960) studied the cultural characteristics of *Morchella* species in submerged culture on standard nutrient medium and found distinct differences between various cultures.

Schmidt (1983) reported that ascospores of *M. esculenta* germinated and developed extensive mycelium within 24 hours, (after one year of storage) at temperatures above 15°C and was slower at lower temperatures, but spore germinated within one week at temperatures of 2°C or above on Malt Extract Agar. This favours Gilbert's (1960) report that *M. esculenta* grows in submerged culture at temperatures above 2.2°C.

Dizbay and Karaboz (1986) raised submerged cultures of various *Morchella* species for microbial protein production. They used various agricultural products and waste material, olive black water, beet molasses, potato and pumpkin as substrates in submerged cultures. The best mycelial development was observed in beet molasses medium. *M. elata* was found to be the most productive species among the *Morchella* strains used in submerged culture experiments. Further, they observed that *M. hortensis* produced a chestnut coloured, fine smelling and nice flavoured very good pallet in beet molasses medium.

Santanna *et al.* (1986) compared the growth of mycelium of *M. crassipes* on synthetic and vinasse medium at 25°C. The growth of mycelium was good in vinasse medium as compared to synthetic medium. The flavour of vinasse medium mycelium was better than the synthetic one. Repin (1901) claims to have obtained fruiting bodies in a cave whose cultures had been established in flower pots.

It is clear, therefore, that cultural studies on morels have been done on aspects whatever the individual concerned considered being important in obtaining the fruiting bodies and that most of the studies have been inconclusive so far as artificial induction of fruiting bodies is concerned.

REVIEW OF INDIAN LITERATURE

Kaul (1977a, b) conducted studies on carbon and nitrogen utilization of morels. He (1977a) studied the carbon utilization in six species of *Morchella* and in *Verpa bohemica*. He used twenty-six different carbon compounds, out of which starch, maltose, fructose, glucose, glycogen and sucrose were good carbon sources for all strains-tested, ribose, sorbose, rhamnose, tartaric acid and malic acid were not utilized by any; and dulcitol, insulin, melinose, and (d-) arabinose gave poor growth in all the cases. Other compounds gave intermediate to poor growth barring a few exceptions, and each strain exhibited an individual pattern of carbon utilization. Kaul (1977b) also studied the nitrogen utilization of nine morel strains, representing six species of *Morchella* and one of *Verpa*, on thirty different nitrogen compounds and observed that asparagine, urea, peptone, glutamic acid, ammonium salts, sodium nitrate and potassium nitrate were good sources of nitrogen for all species, while hydroxylamine-HCl and hydrazine-2HCl were not utilized. Intermediate to poor growth was obtained with the compounds, cysteine, methionine and histidine, glycine, valine, serine, phenylalanine, leucine, isoleucine, and arginine. He also established that individual strains varied in their response to specific compounds.

Kaul (1981) also studied the cultural characteristics of eight morel (*Morchella* spp.) strains representing six *Morchella* species on four media (synthetic and semi-synthetic). He observed different cultural characteristics for different species and strains. Therefore, he concluded that cultural studies have definite use in determining the taxonomic status of species in *Morchella*; however, he suggested that the variations between the strains will have to be investigated further. Kaul (1981) recorded 25°C as the best temperature for the growth of *Morchella* species.

It is clear from the above that various physiological parameters have been investigated by different workers time and again, even if it involves repetition. All these attempts represent genuine desire to understand the behaviour of this fungus in culture so as to achieve cultivation.

Physiological studies conducted on species of morels from Himachal Pradesh to understand the behaviour of *Morchella* spp. in culture are presented in the following pages. No such studies had been conducted earlier on species from Himachal Pradesh and some may appear to be repetitive, but these were required to understand the nutritional requirements and behaviour of these fungi in culture because difference has been noticed in cultural characteristics of not only different species but in different strains as well in the earlier studies by other researchers at different places.

The details of standard methodology for isolation, establishment of cultures and qualitative and quantitative screening of enzymes are given in Annexure–IV a, b .

The mycelial colonies of *Morchella* species were white in colour. The hyphae were observed to be septate, profusely branched, cylindrical, hyaline. Mycelium turned white-brown with age. The hymenium which lies in pits bears numerous asci. These asci are cylindrical/subcylindrical with obtuse apex and each ascus has eight ascospores which are ellipsoid and uniseriate (Plate I a – g).

The mycelial growth of the species was maximum after 10 days of incubation, therefore, in subsequent experiments on nutritional studies, the final data relating to growth of *Morchella* species were recorded only after 10 days of incubation.

OBSERVATIONS AND DISCUSSION

a) Growth of Mycelium on Different Solid Media

Ten solid media were tested for the linear growth of *Morchella* species (Fig. 6.1, Plate I a). The growth was recorded after 10 days of incubation. The results were statistically analyzed. The critical difference was calculated by statistical analysis.

The maximum growth of *M. esculenta* mycelium was in Potato Dextrose Agar followed by Yeastal Potato Dextrose. The minimum growth was found in Pea Extract and Coriander Grain Extract. The maximum growth of mycelium in *M. deliciosa* was also in Potato Dextrose Agar followed by Malt Agar and Yeastal Potato Dextrose and the minimum growth was in the medium Glucose Yeast Agar, Czapeck's Dox Agar and Coriander extract. The maximum growth of mycelium of *M. angusticeps* was in Maize Grain Extract followed by Potato Dextrose Agar and the minimum growth was observed in Pea Extract, Yeastal Potato Dextrose and Wheat Grain Extract. The growth of *M. tibelica* was maximum in Potato Dextrose Agar and minimum in Pea Extract and Yeastal Potato Dextrose Agar.

The maximum growth of mycelium in *M. conica* was in Potato Dextrose Agar and minimum growth was in all other media except Yeastal Potato Dextrose Agar and Malt Yeast Agar. The maximum growth of the mycelium of *M. crassipes* and *Morchella simlensis* sp. nov. was in Glucose Yeast Agar followed by Potato Dextrose Agar and the minimum growth was in Malt Yeast Agar and Pea Extract. The growth of *M. semilibera* mycelium was in Glucose Yeast Agar followed by Potato Dextrose Agar and the minimum growth was observed in Malt Yeast Agar, Wheat Grain Extract and Coriander Extract.

Since the best linear growth of most of the *Morchella* species was recorded in Potato Dextrose Agar, it was used as basal medium in subsequent studies for recording vegetative growth and for maintenance of culture.

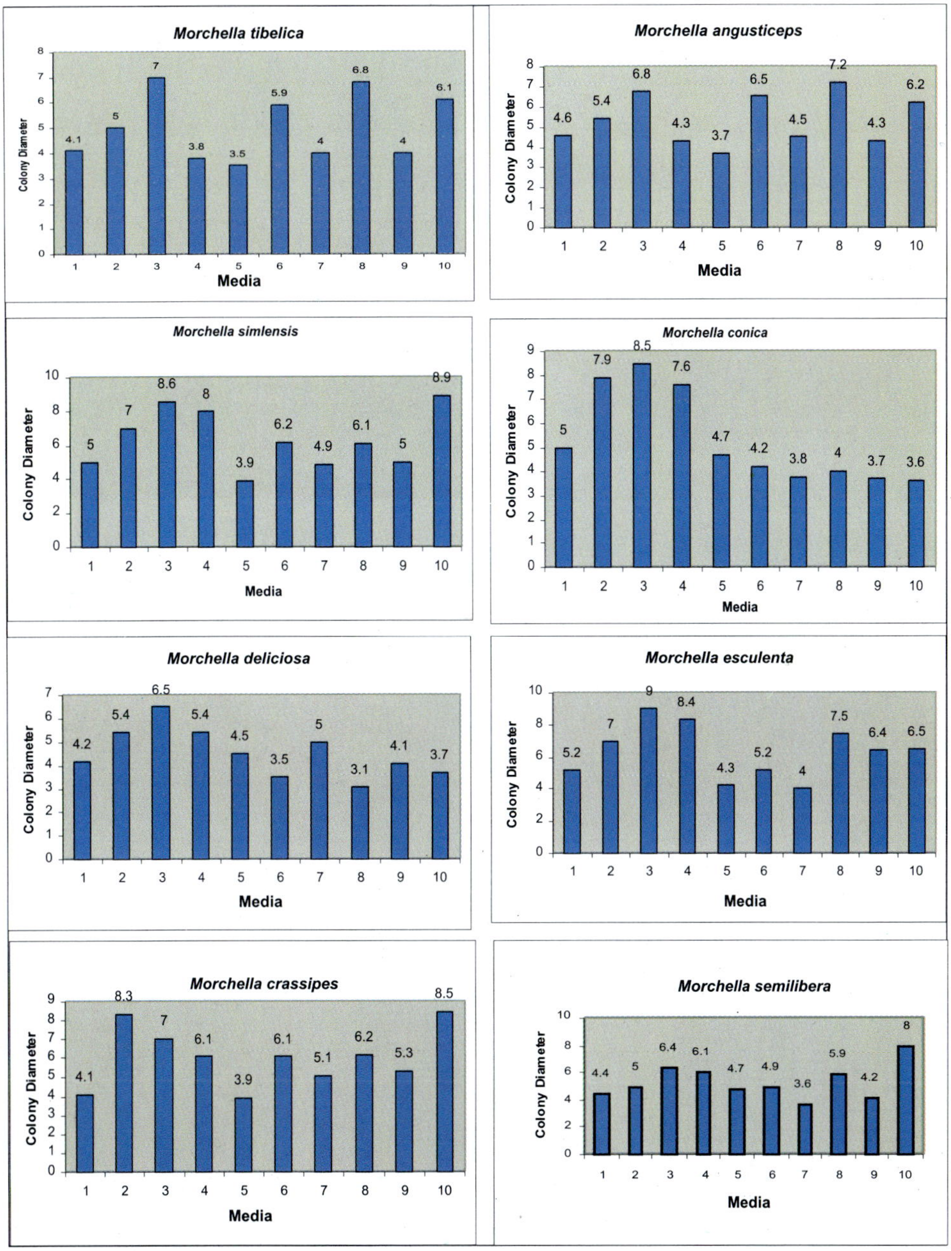

Fig. 6.1 Effect of different solid media on growth of *Morchella* species

Media: 1 – Malt Yeast Agar, 2 – Malt Agar, 3 – Potato Dextrose Agar, 4 – Yeastal Potato Dextrose, 5 – Pea Extract 6 – Czapek's Dox Agar, 7 – Coriander Extract, 8 – Maize Extract, 9 – Wheat Grain Extract, 10 – Glucose Yeast Extract

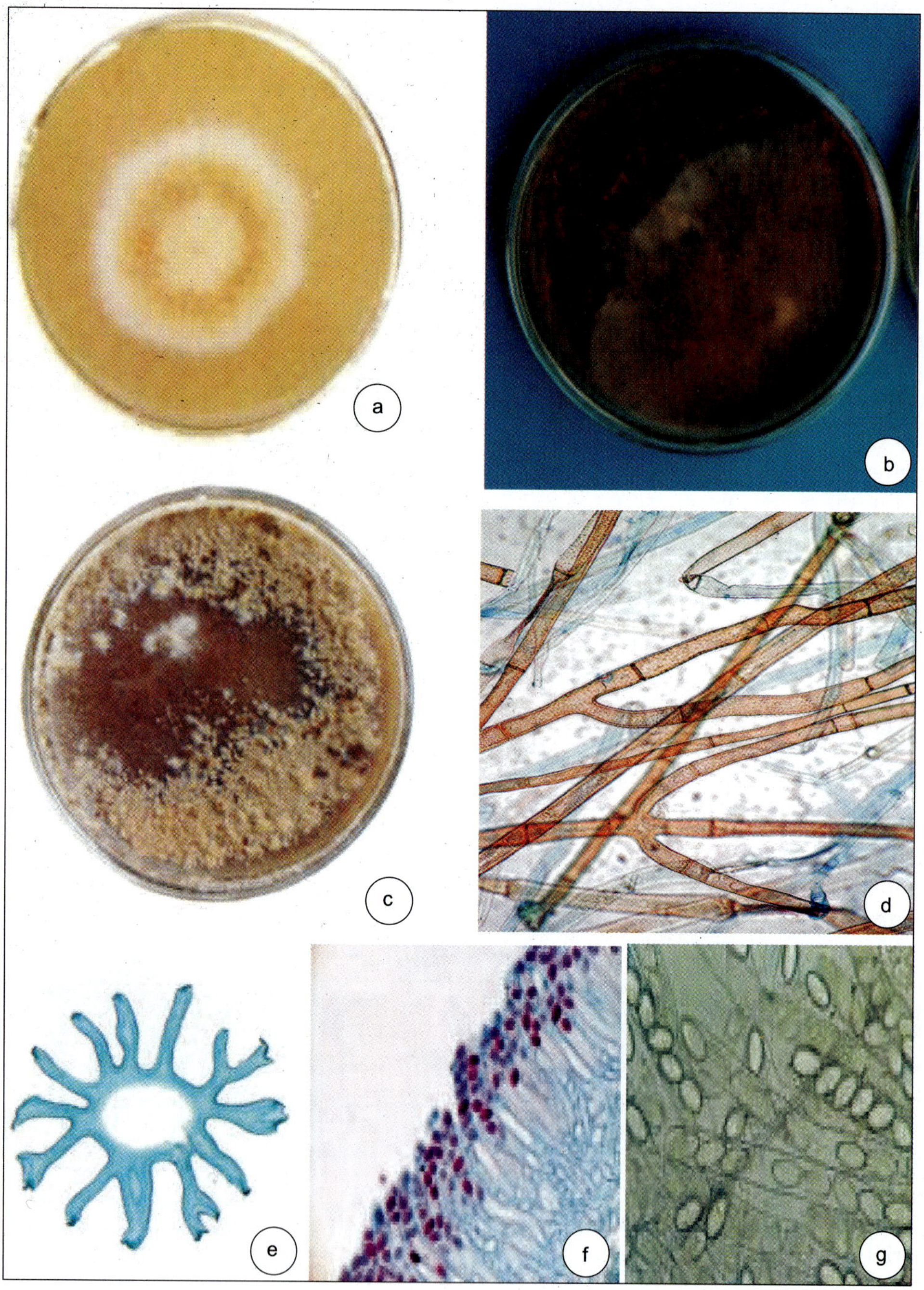

PLATE I (a–g) a, b – Young (a) and old (b) cultures of *Morchella angusticeps*. c – Culture of *Morchella angusticeps* forming sclerotia. d – Mycelial strands of *Morchella angusticeps*. e – A cross-section of sporophore of *Morchella angusticeps*. f & g – Hymenium of *Morchella angusticeps* showing asci, ascospores, and paraphyses.

b) Growth of Mycelium on Different Liquid Media

The vegetative growth in liquid media (Fig. 6.2) was recorded in four synthetic media. The best growth of *M. esculenta* was on Glucose Asparagine medium, whereas it was minimum in Asthana and Hawker solution. In *M. deliciosa*, maximum growth was observed in Glucose Asparagine solution and minimum growth was observed in Dimmick solution. In *M. conica* Glucose Asparagine was the best medium; least growth was in Asthana and Hawker solution. Maximum growth of *M. crassipes* and *Morchella simlensis* sp. nov. mycelium was in Glucose Asparagine solution, and minimum in Dimmick's solution. The maximum growth of *M. angusticeps* and *M. tibelica* was in Asthana and Hawker solution and minimum growth in Czapek's solution. In *M. semilibera* the maximum growth was in Glucose Asparagine and minimum in Potato Dextrose Agar.

The effect of temperature on linear growth of mycelium in different *Morchella* species revealed that the temperature between 15°C and 25°C is optimum for the growth of mycelium and minimum growth was observed at 5°C (Fig. 6.3).

The effect of pH on the linear growth of mycelium revealed that the maximum growth of mycelium in the species of *Morchella* is at neutral pH (i.e. 7.0) followed by basic pH (i.e. 7.5). Minimum growth was observed at pH 5.0, 4.0, 3.0 in all the species (Fig. 6.4).

The effect of light and darkness was also observed on the growth of mycelium in different species of *Morchella*. It was observed that mycelial growth in dark was better than in light (Fig. 6.5). All the solid media tried supported good growth of mycelium. However, the best growth of *M. esculenta M. conica, M. tibelica* and *M. deliciosa* was recorded on Potato Dextrose Agar media and that of *M. crassipes, Morchella simlensis* sp. nov. and *M. semilibera* on Glucose Yeast Agar. Best mycelium growth of *M. angusticeps* was recorded on Maize Extract.

The exact reason for this difference of growth in dark and light could not be ascertained, but it may be due to the inhibitory effect of light on assimilation of growth substances and growth producing hormones.

These results corroborate the previous findings of different workers. It is also clear that species of *Morchella* do not behave much differently in culture than other cultivated mushrooms. But there is some aspect which is still intriguing, and it is that which is required to be understood for triggering fructification development.

(II) ANALYSIS OF ENZYMATIC POTENTIAL

INTRODUCTION

Enzymes are water-soluble, colloidal, organic macromolecules which are wholly and partially proteinaceous in nature, biological in origin and capable of catalyzing specific biochemical reactions under specific conditions of temperature and pH.

A variety of yeast, fungi and bacteria degrade a range of complex polymeric substances in soil by secretion of some extracellular enzymes. Fungi are the potent source of enzymes that are preferred over the enzymes produced by animals and plants (Fogarty and Kelly, 1979). The fungi produce extracellular enzymes that break down organic compounds present in the colloidal material surroundings the roots.

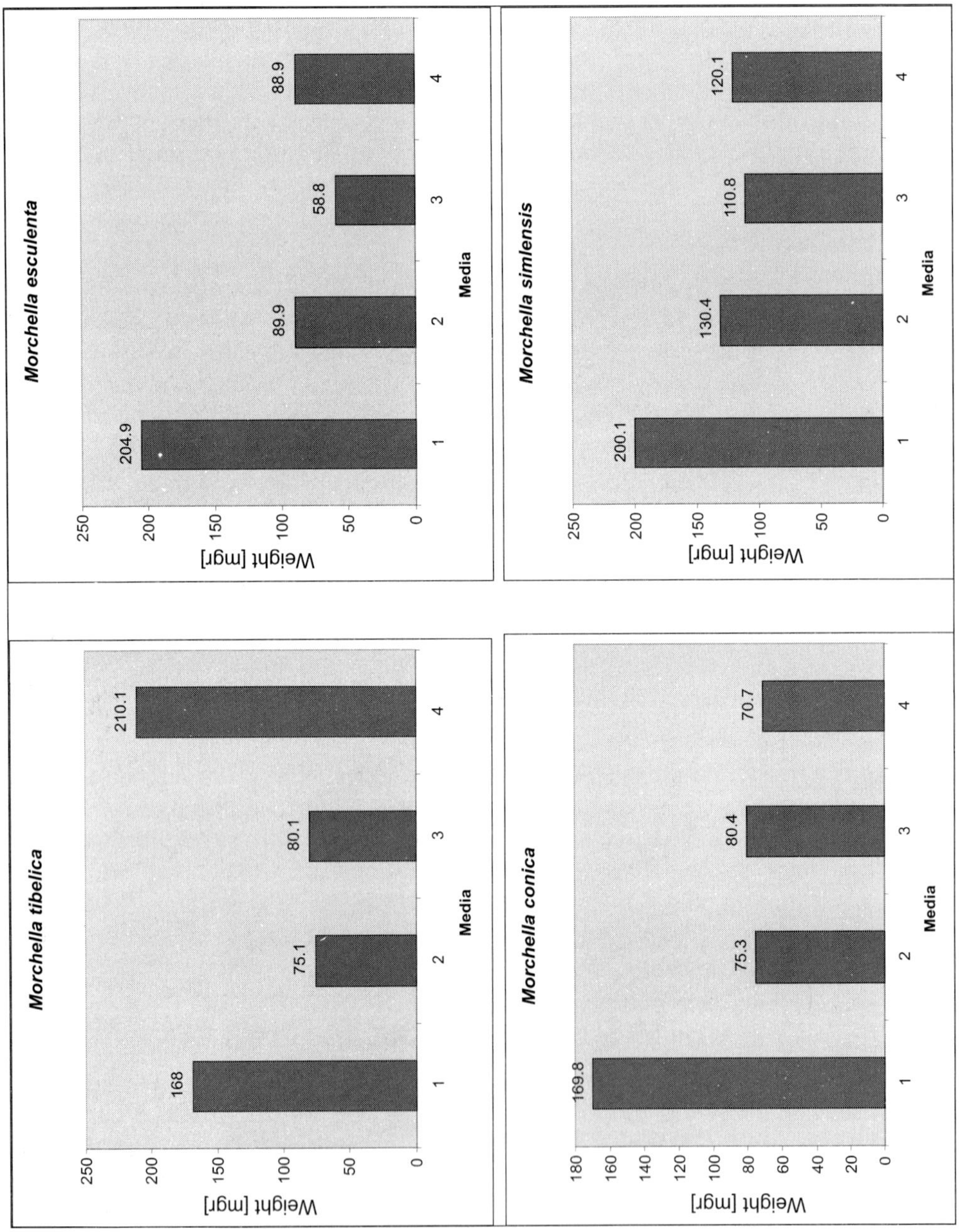

Morchella esculenta
204.9
89.9
58.8
88.9
Morchella simlensis
200.1
130.4
110.8
120.1
Morchella tibelica
168
75.1
80.1
210.1
Morchella conica
169.8
75.3
80.4
70.7
Weight [mgr]
Media

(Fig. Contd.)

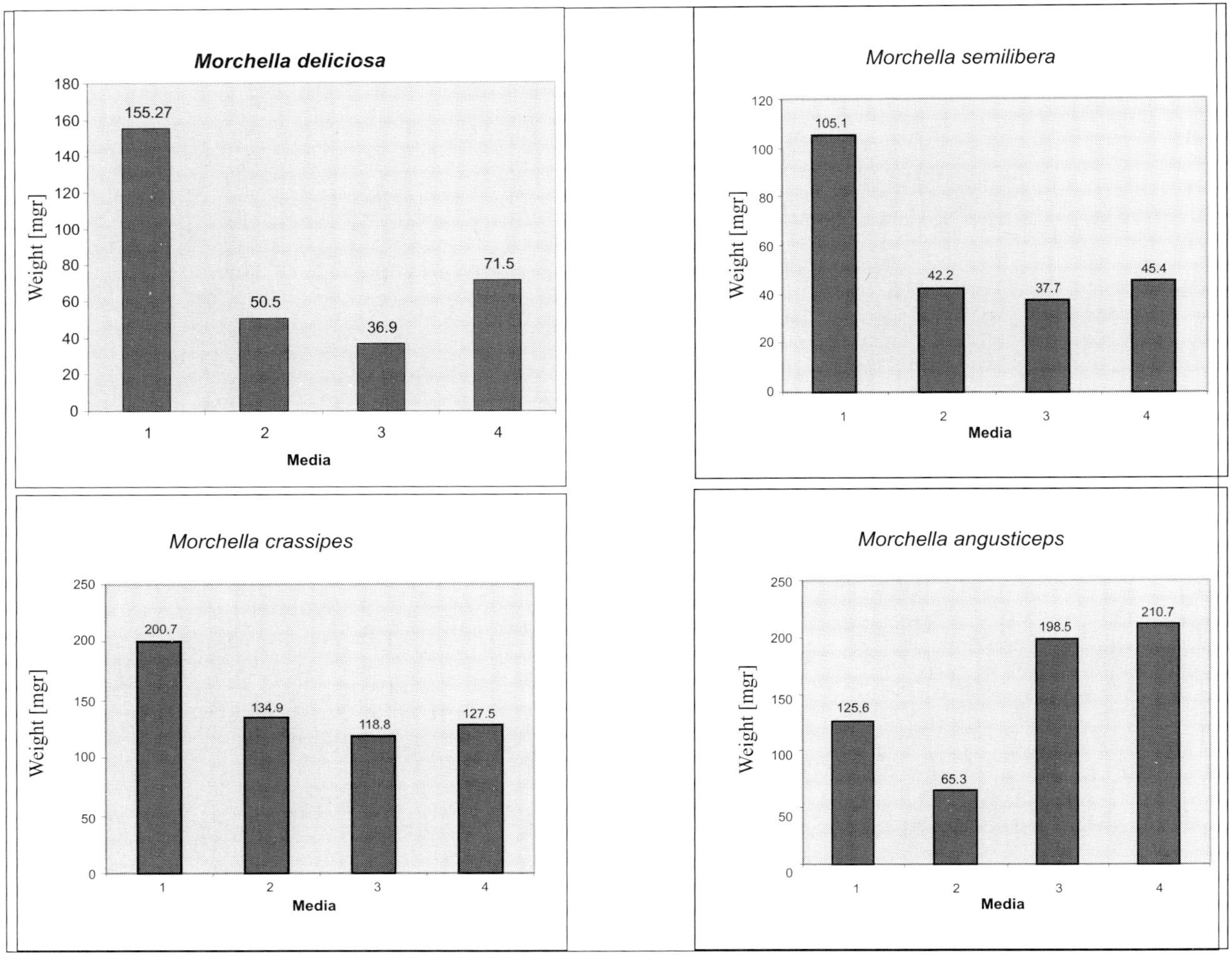

Fig. 6.2 Effect of different liquid media on growth of *Morchella* species
1 – Glucose Asparagine solution; 2 – Czapeck's solution; 3 – Dimmick's solution; 4 – Asthana and Hawker solution

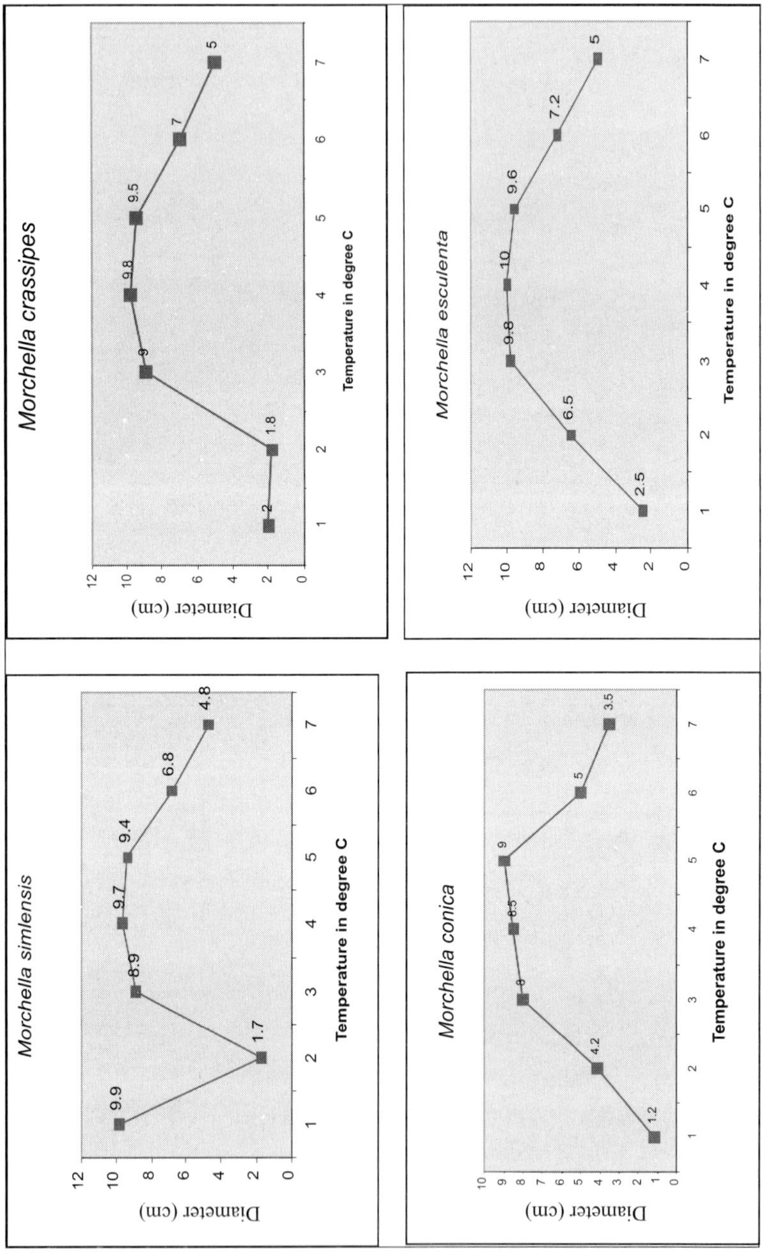
Morchella crassipes
Morchella esculenta
Morchella simlensis
Morchella conica
Diameter (cm)
Temperature in degree C

(Fig. Contd.)

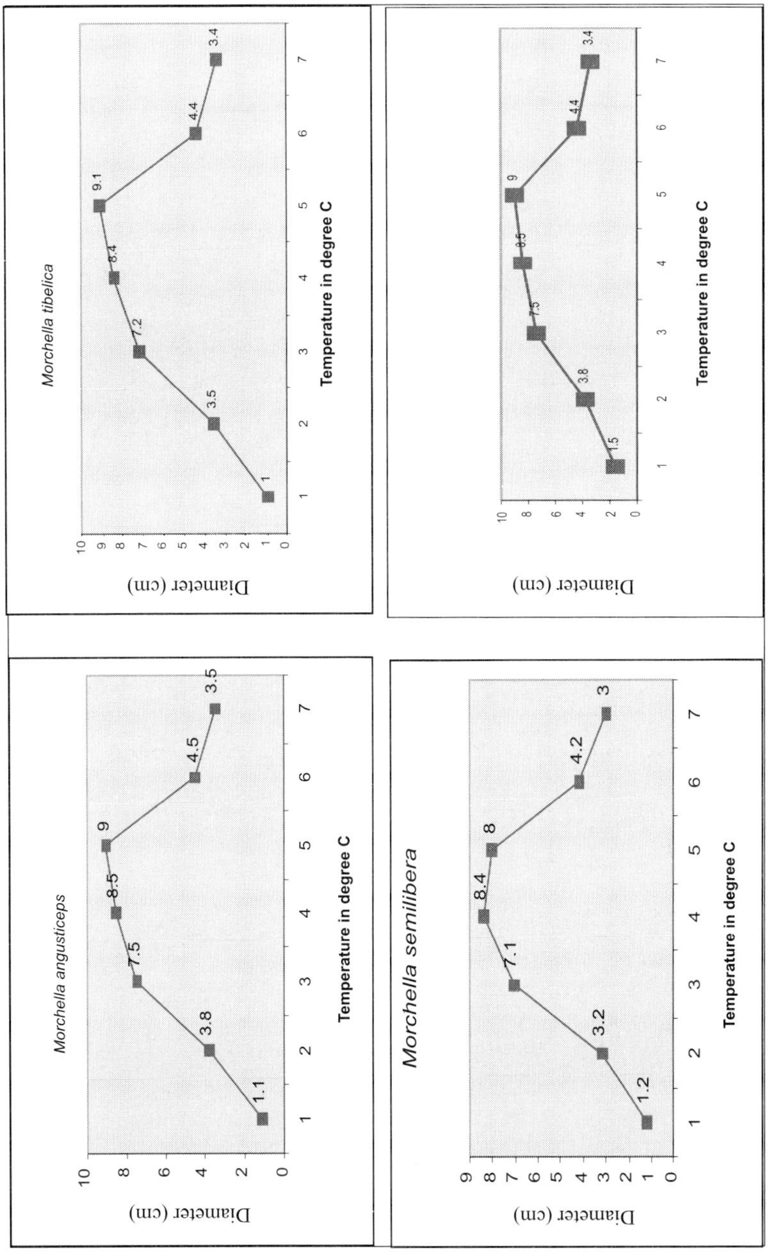

Fig. 6.3 Effect of temperature on growth of *Morchella* species

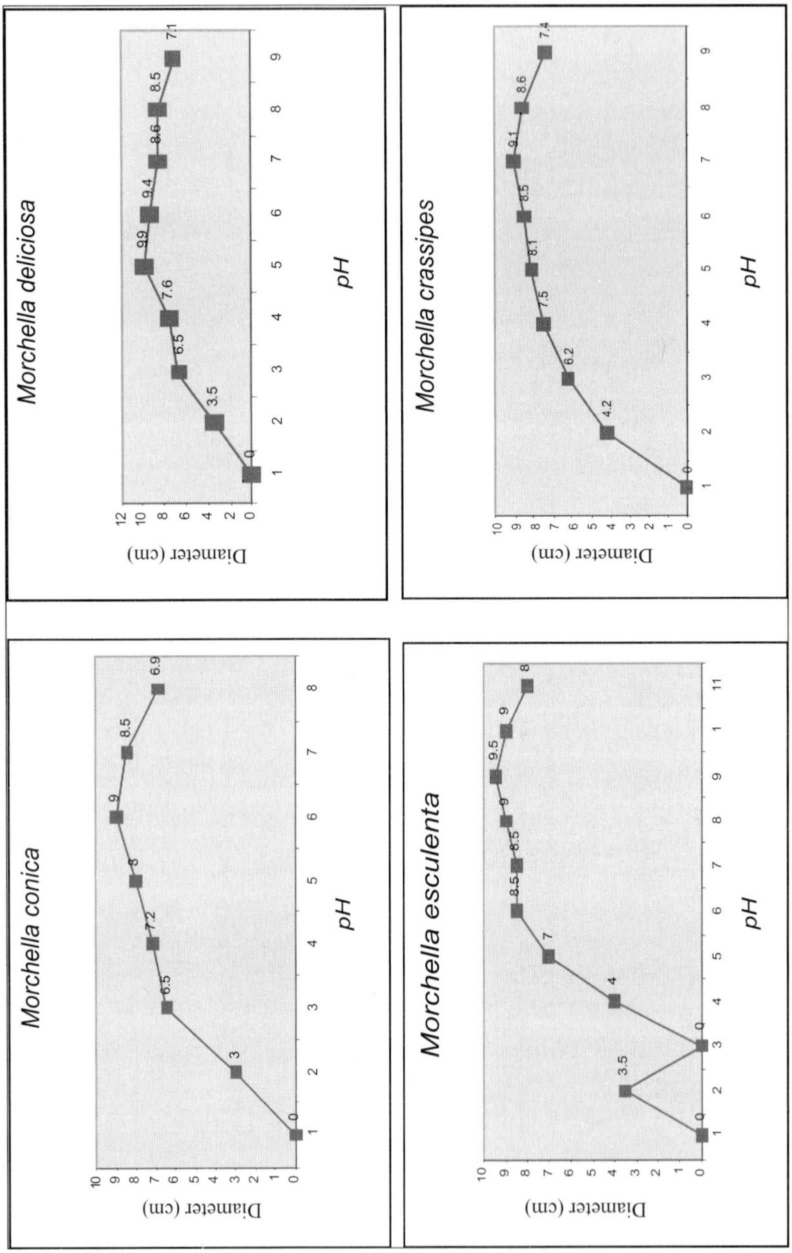
Morchella deliciosa
Diameter (cm)
pH
Morchella crassipes
Diameter (cm)
pH
Morchella conica
Diameter (cm)
pH
Morchella esculenta
Diameter (cm)
pH

(Fig. Contd.)

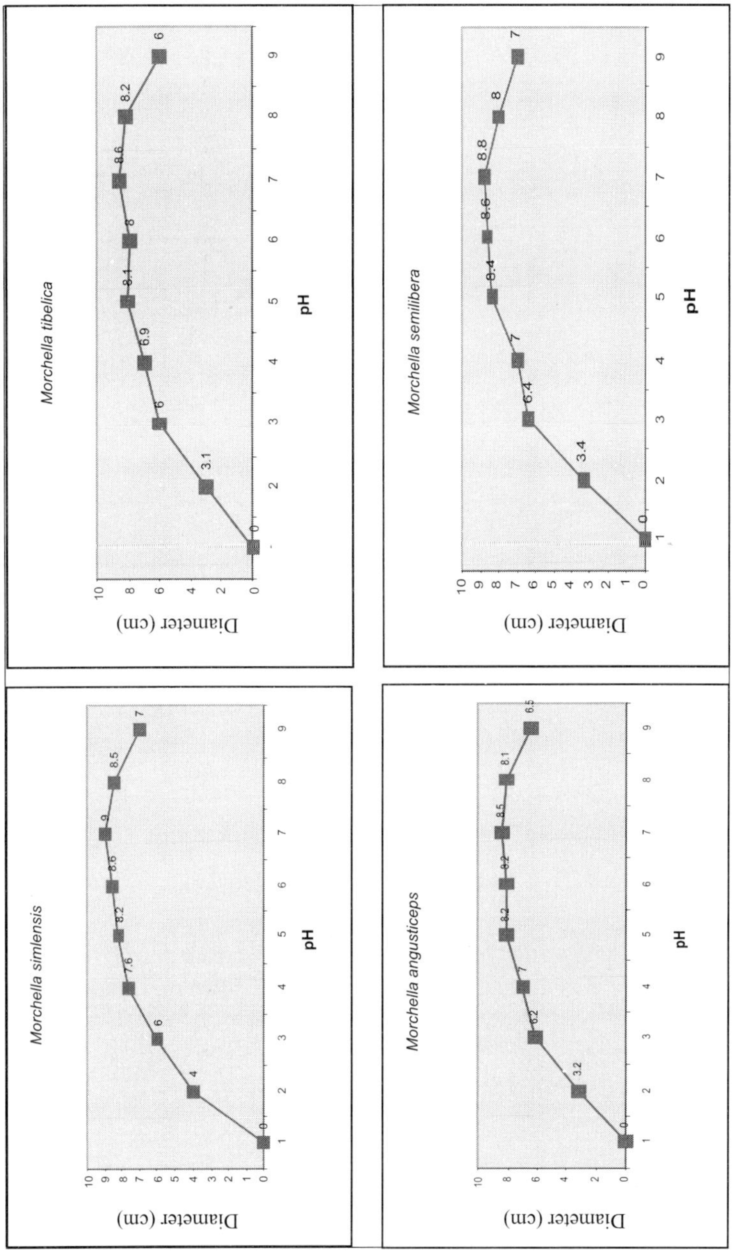

Fig. 6.4 Effect of pH on growth of *Morchella* species in solid medium

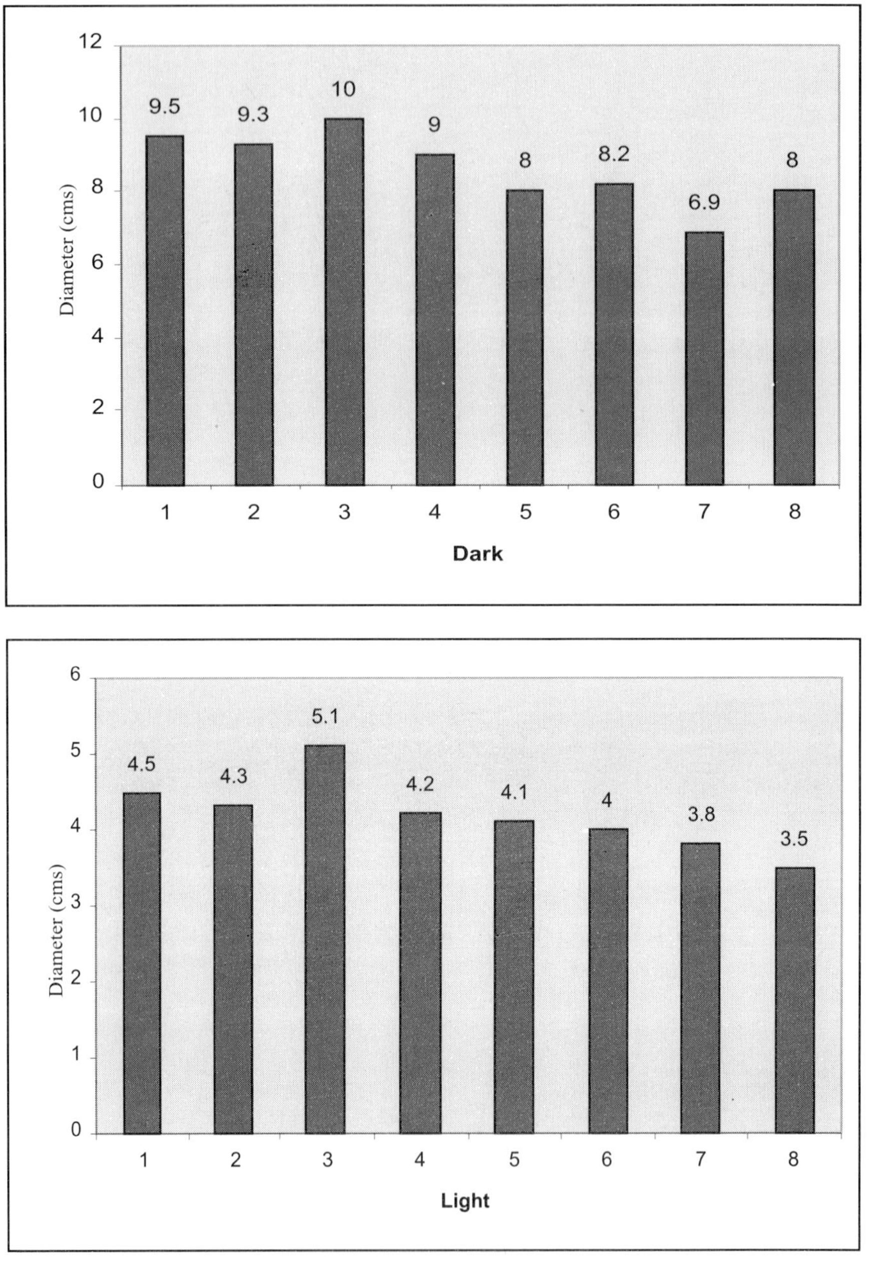

1 – *M. crassipes*, 2 – *M. simlensis* sp.nov., 3 – *M. esculenta*, 4 – *M. conica*, 5 – *M. tibelica*, 6 – *M. angusticeps*, 7 – *M. deliciosa*, 8 – *M. semilibera*

Fig. 6.5 Effect of light and dark on growth of *Morchella species*

Macro fungi (mushrooms) produce a wide range of extracellular enzymes and these extracellular enzymes enable them to degrade complex organic matter into soluble substances, which can then be absorbed by the mushrooms for the purpose of nutrition actively or passively. Therefore, the growth and fruiting of any individual mushroom species on a particular waste material will hence depend largely upon the ability of that mushroom to produce enzymes, which are essential to degrade the major components of the waste particular (or its growth substrate) and thereafter absorb it as food. Therefore, fungi obtain much of their nutrition by extracellular enzymes.

In this way, the capability of the mushroom to grow on a particular type of substrate lies in its ability to degrade it, which in turn is decided by the repertoire of enzymes the mushroom possess, i.e. the set of enzymes a mushroom species can produce defines the set of substrates on which it can grow, while the comparative intensity of the activities of different enzymes decides its preference for a substrate over others (Cavazzoni and Manzoni, 1994 and Rai *et al.*, 1990).

Interest in the cultivation of morels has increased as a result of recent biochemical and genetic studies (Ower, 1982 and Ower *et al.*, 1986). But besides precise controlled conditions, substrate selection for possible domestication of *Morchella* species remains a challenge for mycologists. So, present studies were conducted to find out the possible role of production of extracellular enzymes by *Morchella* species in degrading substrates, which can provide vital clues for the selection of substrates for its cultivation trials. The following parameters were investigated:

- To check the qualitative screening of some extracellular enzymes, viz. cellulase, amylase, pectinase, lipase and urease in *Morchella* species.
- To screen the quantitative production of some commercial enzymes: amylase, protease and cellulase from vegetative mycelium of *Morchella* species.

REVIEW OF LITERATURE

In the hyphae of mushrooms, the secretory enzymes are synthesized and accumulated to digest the complicated substances that surround them and break them down to the simple molecules. Common fungal exoenzymes are diverse and include amylase, cellulase, protease, xylanase, ligninase, oxidases, etc.

The taxonomic confusions in the genus *Morchella* have been tried to be resolved using methodologies beyond morphology. Since the set of enzymes of a mushroom species defines the set of substrate to be used and the varied culture conditions do not affect the repertoire of enzymes, this could be used as genetic marker in the species distinction (Kulkarni and Kamerath, 1989). Morels may be placed, in general, in three broad groups based on their gross morphology: black morels (*M. angusticeps, M. conica,* and *M. elata*), yellow morels (*M. esculenta, M. crassipes,* and *M. deliciosa*) and semi free-capped morels—*M. semilibera* (Bunyard *et al.*, 1994). Gessner *et al.* (1987) using enzyme electrophoresis classified *M. deliciosa* and *M. esculenta* as two polymorphic populations of the same species. Ower (1982) and Ower *et al.* (1986) observed that accomplishment of small-scale laboratory production and secretion of extracellular enzymes by *Morchella esculenta* and *M. angusticeps* enhances their possibilities of domestication.

The high production of extracellular lignocellulytic enzymes in *Morchella esculenta* was reported by Singh *et al.* (2001). They assessed two isolates of *Morchella esculenta* for the production of *in vitro* extracellular endoglucanase, β-glucosidase, xylanase, laminariase, laccase and polyphenol oxidase enzymes activities on the wheat straw infused liquid medium with or without wheat flour. Both the isolates of *Morchella esculenta* produced appreciable activities of all the six extracellular enzymes studied, which indicates that *Morchella esculenta* is capable of degrading and utilizing cellulose, hemicellulose and lignin.

The role of enzymes in initiating sexual cycle in different species of *Morchella* was studied by Kamal *et al.* (2004). They studied the variation in the production of seven different enzymes, viz. cellulase, xylanase, β-glucosidase, laccase, polyphenol oxidase and peroxidase during the major life cycle stages, namely vegetative mycelial stage, dormant sclerotia stage and sexual ascocarp formation stage in three different *Morchella* spp., viz. *M. esculenta, M. angusticeps* and *M. conica*. In general, they observed that mycelial extracts exhibited lesser activities of all the seven enzymes and the dormant sclerotial stage, which later on reduces gradually from sclerotial to fruit body maturation stage. They also suggested that sexual cycle may be triggered by manipulating enzymatic activities of vegetative sclerotial stage. Mayr (1982) observed that low production of laccase and polyphenol oxidase in *M. conica* is attributed to its mycorrhizal nature, and it is ectomycorrhizically associated with the roots of *Pinus sylvestris*.

Tiwari *et al.* (2004) compared the enzyme profile of eight species of *Morchella*, viz. *M. esculenta, M. angusticeps, M. conica, M. crassipes, M. spongicola, M. vulgaris* and related genera, viz. *Mitrophora semilibera*, and *Verpa conica* for seven extracellular enzymes. The species of genus *Morchella* produced significantly higher quantities of cellulase and xylanase in comparison to the related genera *Mitrophora* and *Verpa*. Black morels produced significantly higher amounts of laccase in comparison to yellow morels. This study confirmed higher extracellular polymorphism at interspecific than intraspecific levels. The polymorphism in enzyme profiles could distinguish yellow morels, black morels and related genera from each other.

In the present work the extracellular enzyme polymorphism at inter and intraspecific levels in different species of morels in Himachal Himalaya is investigated.

OBSERVATIONS AND DISCUSSION

1) Qualitative Screening of Different Extracellular Enzymes by *Morchella* spp.

Four *Morchella* species, viz. *M. angusticeps, M. conica, M. deliciosa* and *M. esculenta* were screened for the production of five extracellular enzymes. All the four species were positive for the production of cellulase, pectinase, amylase and urease but were negative for the production of lipase. The qualitative screening of extracellular enzyme profile has been presented in Table 6.1.

When all the species were compared, *M. angusticeps* and *M. conica* exhibited lesser production of cellulase and amylases than *M. esculenta* and *M. deliciosa,* while the production of pectinase and urease was same in all the investigated species. All the species showed negative results for lipase. From these investigations, it was observed that *M. esculenta* and

Table 6.1 Qualitative screening of vegetative mycelium of *Morchella* spp. for different extracellular enzymes

Sr. No.	*Enzyme tests*	*M. angusticeps*	*M. conica*	*M. esculenta*	*M.deliciosa*
1.	Cellulase	+	++	+	++
2.	Lipase	–	–	–	–
3.	Pectinase	++	++	++	++
4.	Amylase	+	++	+	++
5.	Urease	++	++	++	++

M. deliciosa were able to utilize different carbon sources and produced extracellular enzymes better than *M. angusticeps* and *M. conica.* The production of all of these test enzymes was recorded to be higher in *M. esculenta* and *M. deliciosa* as compared to *M. angusticeps* and *M. conica.* Thus, the observations again inferred that yellow morels (*M. esculenta* and *M. deliciosa*) showed higher production levels for different extracellular enzymes.

No such work on qualitative screening of extracellular enzymes was carried out in *Morchella* spp. earlier. But similar studies were carried out on the qualitative screening of extracellular enzymes in *Cantharellus cibarius,* an ectomycorrhizal mushroom. The mycelium of *C. cibarius* was investigated for five enzymes, viz. cellulase, pectinase, amylase, urease and lipase and showed positive results for all of them (Sharma, 2002).

2) Quantitative Screening of Vegetative Mycelium of *Morchella* Species for Different Extracellular Enzymes

The vegetative mycelium of four different *Morchella* species, viz. *Morchella angusticeps, M. conica, M. esculenta,* and *M. deliciosa* was screened for the production of three different extracellular enzymes: amylase, cellulase, and protease. The activity of these respective enzymes was checked in supernatant portion. The enzyme profile of four *Morchella* spp. has been presented in Table 6.2.

The amylase activity was found to be 0.04 units in *M. angusticeps;* 0.03 units in *M. conica;* 0.05 units in *M. esculenta* and 0.05 units in *M. deliciosa.* So, the maximum activity of amylase was observed in *M. esculenta* and *M. deliciosa.*

Table 6.2 Quantitative screening of the vegetative mycelium of *Morchella* species for different extracellular enzymes

Sr. No.	*Morchella spp.*	*Amylase (I.U.)*	*Cellulase (I.U.)*	*Protease (I.U.)*
1.	*Morchella angusticeps*	0.04	0.001	0.073
2.	*Morchella conica*	0.03	0.008	0.075
3.	*Morchella deliciosa*	0.05	0.03	0.09
4.	*Morchella esculenta*	0.05	0.02	0.14

I.U. = International Units

Incubation period = 7-12 days.

The protease activity was 0.073 units in *Morchella angusticeps*; 0.075 units in *M. conica;* 0.09 units in *M. deliciosa* and 0.14 units in *M. esculenta.* The maximum activity of protease was observed in *M. esculenta* and *M. deliciosa* than *M. angusticeps* and *M. conica.*

In the enzyme assay of cellulase in culture, supernatant of various *Morchella* spp. was observed. The activity revealed 0.001 units in *M. angusticeps*; 0.008 units in *M. conica;* 0.03 units in *M. esculenta* and 0.02 units in *M. deliciosa.* So, again the maximum activity of cellulase was observed in *M. deliciosa* than *M. esculenta* and rests all species.

After screening all the three selected enzymes, it was found that the maximum activity was observed in protease, followed by cellulase and amylase. *M. esculenta* and *M. deliciosa* (yellow morels) showed maximum activity of extracellular enzymes than *M. angusticeps* and *M. conica* (black morels). So, the enzyme profiles also revealed the interspecific polymorphism in *Morchella* spp. The production of all these enzymes shows that *Morchella* spp. are capable of degrading and utilizing substrates such as cellulose, starch, urea, pectins and protein efficiently, and this can help the mycologists to opt such substrates for the domestication of the species.

Certain other studies were also carried out previously to check the extracellular enzyme screening in different *Morchella* species. Kulkarni and Kamerath (1989) separated yellow and black morels using isozyme electrophoresis of NAD-glutamate dehydrogenase and glucose phosphate isomerase. Royse and May (1990) also separated yellow morels from black morels by analyzing twelve different enzymes.

In a similar study, Singh *et al.* (2001) recorded the production of lignocellulotic enzymes by *Morchella esculenta* isolates and noticed that *M. esculenta* is capable of degrading and utilizing cellulose, hemicellulose and lignin. Kamal *et al.* (2004) observed similar types of observations while screening different extracellular enzyme production by *Morchella* spp. viz. *M. conica, M. angusticeps,* and *M. esculenta.* They observed seven different enzymes, viz. cellulase, xylanase, β-glucosidase, laccase, polyphenol, oxidase and peroxidase which were produced by three *Morchella* spp. and suggested that morels are capable of degrading substrates such as lignin, cellulose and hemicellulose efficiently.

Tiwari *et al.* (2004) reported the enzyme profile of eight species of *Morchella* and the related genera for seven different extracellular enzymes. They observed that cellulase production was higher in yellow morels (*M. esculenta, M. crassipes* and *M. deliciosa*) than black morels (*M. angusticeps* and *M. conica*) and the related genera.

Thus, on the basis of the enzyme profiles developed during present investigations, yellow and black morels can be separated from each other. The results of the present study confirm that polymorphism at interspecific levels are higher in respect of all extracellular enzymes. With the production of all the selected three enzymes (cellulase, amylase and protease), it was inferred that *Morchella* spp. are capable of degrading and utilizing various specific substrates, viz. cellulose, starch, urea, pectins and proteins more efficiently.

CHAPTER

7 Nutritive* Composition

CHAPTER OUTLINE

- Introduction
- Review of Literature
- Observations and Discussion

INTRODUCTION

Mushrooms, though mostly regarded a delicacy, are realistically a food that can contribute significantly in relieving protein deficiency. Mushrooms are low calorific but protein rich and hence a good source of protein for human consumption (Kurtzman, 1975). Normally, mushrooms contain 20-40 per cent protein on a dry weight basis and thus surpass many foods including milk in terms of protein content. Most species contain all the essential amino acids and most of them are found in about the same proportion as in egg. Mushrooms, like yeasts, are also rich in vitamins, with the exception of thiamine.

There are approximately 2000 edible fungi. Out of these only about 30-40 species have been cultivated so far. However, in addition to the few cultivated ones, there are many species of edible fungi which are still collected from the wild. These form an important item of man's culinary habits and in certain areas constitute an item of commerce as well.

Species of *Morchella,* commonly called "morels", are still mainly collected from the wild. These have an excellent flavour and are reputed to be superior to that of other mushrooms. Not only morels but even their mycelia have been reported to have the same nutritive value (Cochrane, 1958; Robinson and Davidson, 1959; Singer 1961; Hayes and Hadded, 1976). In this chapter the nutritive analysis of six species of *Morchella* from Himachal Pradesh is presented.

In fact, all the six species have never been analyzed simultaneously for nutritive components in India earlier. The results of chemical analysis are presented in Tables 7.1-7.3. The procedures followed for chemical analysis of different components are given with material and methods.

REVIEW OF LITERATURE

A sizeable number of research papers have appeared during the past on the nutritive value of different mushrooms. Some of these concern a systematic analysis of almost all the contents whereas others report presence of various important components only. In most instances the tendency has been to analyze components and then to interpret and compare them with those of other traditional sources of diet such as cereals, eggs, meat and legumes. Hence, different connotations have been used to emphasize their influence and to equate them with the available food sources, e.g., "vegetarian meat", "vegetable beef steak", etc., while others consider them of minor nutritional significance (Chang and Hayes, 1978).

In fact, few definitive studies conducted by competent nutritionists are available and hence little is known about the true nutritive value of mushrooms. Most of the data on the nutritional contribution of mushrooms to human diet have been derived from empirical determinations of mushroom composition, especially of the protein content (Chang and Hayes, 1978).

---

*Nutraceutical potential of morels is discussed in Chapter 8. There may be some repetition of constituents analyzed in these two chapters, but in chapter on nutraceuticals, the morels have been grouped into two groups, whereas in this chapter analysis of components of each species has been done and presented separately. This chapter forms the basis of Chapter 11.

The comments of Quackenbush *et al.* (1935) that "Although most of the interest has been directed to the protein fraction, little can as yet be said with certainty regarding the actual protein present and their digestive and nutritive properties", hold true even today. Recent researches have, however, shown mushrooms to be a good source of vegetable protein. Another advantage of mushroom is that the cultivated mushrooms can produce more protein per unit area of land than any other form of agriculture, and since the economics of mushroom production is labour intensive, it can be helpful in solving protein deficiency problems in developing countries where the labour is comparatively cheap. Further, mushrooms can be used to supplement the protein component in the third world countries where majority of the people eat protein-deficient diets (Rehman Espig, 1976).

Food value of edible mushrooms has been analyzed by a number of researchers notably, Mendel, 1898; Quackenbush *et al.*, 1935; Rambsbottom, 1935; Gilbert and Robinson, 1957; Singer, 1961; Atkinson, 1961; Gray, 1970; Kurtzman, 1975; Hayes, 1975; Hayes and Hadded, 1976 and Crisan and Sands, 1978. Contradictory views, however, have been expressed about the nutritive quality of several mushroom species, which is not unusual because the age of sporophores, mycelia, nature of substrate and fungal strains sometimes affect the nutritive quality of a species.

Analysis of crude mushroom protein suggests that only 34-89 per cent of the protein (N 6.25) is digestible (Lintzel, 1943; Morner, 1986; Saltet, 1885). Other studies indicate a probable digestibility of 60-70 per cent (Fitzpatrick *et al.*, 1946; Gilbert and Robinson, 1957; Skinner *et al.*,1933). This reduced coefficient of digestibility can be explained by the fact that mushrooms contain a significant amount of non-protein nitrogen in their chitinous cell walls, nitrogen which would be calculated as crude protein after standard nitrogen analysis. Protein content may vary from as low as 4-9 per cent for species of *Auricularia* to as high as 44 per cent for *Agaricus* species. Based on dry weight, *A. bisporus* strains alone indicate a range of 24-44 per cent crude protein. *Lentinus edodes* has been reported to have 17.5 per cent crude protein on dry weight basis (Sugimori, 1971). Bano *et al.* (1971) reported 28.5 per cent protein in *Volvariella diplasia* (Berk and Br.) Sing.

The protein content in different species of *Pleurotus* has been recorded to vary from 8.9-38.7 per cent (Bano *et al.* 1963, 1981; Ranged, 1977; Khanna and Garcha 1981; Adriano and Curz, 1933, and Mendel, 1898). Protein content of *Sparassis crispa* was estimated to be 19.375 (Sharma, 1987) and that of *Octaviania densa* (Rodow.) Cunn. as 14.375 per cent (Singh, 1988).

Kaul (1978) recorded 31.06 to 36.06 per cent protein content in morel sporophores. He recorded mycelial protein content to be 20.31-26.87 per cent. He observed that mycelial protein content, however, depends upon the nitrogen content of the medium, and increase from 22 to 37 per cent has been obtained in *M. hortensis* by Chang in C-N ratio (vide Litchfield, 1964). Jandaik (1979) also observed 20-35 per cent protein content in the carpophores of *M. angusticeps.*

Mushroom proteins contain most of the lysine and tryptophan but are deficient in sulphur-containing amino acids. Since proteins are made up of over twenty different amino acids in varying amounts, they are qualitatively different. The human body can convert some of these into others, but there are nine of them that the body cannot make, i.e. lysine, methionine, tryptophan, threonine, valine, leucine, isoleucine, cystine, and phenylalanine, (Chang, 1982). The food products of animal origin always give a better balance and higher quantity of protein than plant food, which often lacks some of the important amino acids, e.g. cereal grains have too little of the essential amino acid lysine, while legumes usually lack the essential amino acid methionine and tryptophan. But the mushroom protein supplies all the nine essential amino acids to man (Chang, 1980). In most of the cultivated and wild edible species of mushrooms, a total of 18 amino acids, namely isoleucine, leucine, lysine, methionine, cystine, phenylalanine, tyrosine, threonine, tryptophan, valine, arginine, histidine, alanine, aspartic acid, glutamic acid, glycine proline, serine are present (FAO, 1970; 1972; 1973 Kalberer and Kunsch, 1974; Lunven *et al.*, 1972; Morner, 1886; Seelkopf and Schuster, 1957; Sugimori *et al.*, 1971, Bano *et al.*, 1963, 1964, 1971).

The total amino acid content, 25-40% is comprised of essential amino acids within a range of 10-50% in mushrooms (Le Rux and Danglot, 1972; Maggioni and Renosto, 1970). Approximately 25-35 per cent of the total amino acids occur as free amino acids, the remainder being combined in the protein. In addition to the common acids and amides, less common or unusual amino acids and related nitrogenous compounds have been detected in various mushroom species.

McKeller and Kohrman (1975) studied the amino acid composition of the sporophores of *M. crassipes, M. deliciosa, M. esculenta* and *Morchella* spp. and they reported arginine, aspartric acid, cystine, glutamic acid, methionine, phenylalanine, proline, serine, threonine, tryptophan, tyrosine, and valine from the sporophores of *M. angusticeps, M. concia, M. crassipes* and *Verpa bohemica*. They also recorded that all the species of *Morchella* possess lysine and tryptophan-amino acids in which vegetables are deficient.

Kaul (1978) reported the following mineral constituent in the fruiting bodies of some morel species. Ash, HCl-insoluble, sesquioxide, Fe_2O_3, CaO, P_2O_5, K_2O, MgO, ranged 11.87 - 17.77%, 2.059 - 5.771%, 1.008 - 2.456%, and 0.3178 - 0.7997%, 3.121-3.475%, 3138 - 3.954%, and 0.3178 - 0.4821%, respectively in the morel fruit bodies of *M. conica, M. esculenta. M. deliciosa, M. angusticeps, M. hybrida* and *Verpa bohemica.*

In the mycelia of *M. angusticeps, M. esculenta* and *V. bohemica*, ash, HCl- insoluble, Sesquioxide, Fe_2O_3, CaO, P_2O_5, K_2O, MgO range 4.58 - 7.18%, 1.007 -1.578%, 0.628 - 1.237%, 0.371 - 0.751%, 0.231 - 0.35%, 1,125 - 7.831, 0.923 - 1.914% and 0.078 - 0.095%, respectively. It is clear that in general sporophores had higher mineral content than the mycelia.

Fresh mushrooms contain relatively large amounts of carbohydrates (3.28%) and fibre (3.32%). The carbohydrate content may consist of a large variety of compounds. Polymeric carbohydrates include glycogen, which serves as an energy storage compound comparable to starch in higher plants (McConell and Esselen, 1947), and chitin ("fungus cellulose"), a polymer of N-acetyl glycosamine, the structural component of fungus cell wall. Mushrooms are relatively good source of fat, phosphorus, iron, thiamine (B_1) riboflavin (B_2), and niacin, They are low in calories, carbohydrates and calcium, but high in ascorbic acid (Vitamin C) in *Agaricus* and ergosterine (Vitamin D) in *Lentinus* and *Volvariella.* All mushrooms, however, are devoid of vitamin activity. Kaul (1978) recorded high phosphorus, potassium and iron but lesser amount of calcium in sporophores of *Morchella.*

Mushrooms as well as their mycelia have sufficient nutritive value to serve as a supplementary source of proteins, B-vitamins and minerals (Cochrane, 1958; Robinson and Davidson, 1959; Singer, 1961; and Hayes and Hadded, 1976). Russer *et al.* (1958) reported morel mycelium to be good source of vitamins especially, the B-group vitamins. Robinson and Davidson (1959) observed that among the different species of *Morchella, M. crassipes* gave the best flavour.

Litchfield *et al.* (1963) in quantitative amino acid evaluation of morel mycelia found them deficient in cystine and methionine. Janardhan and Hussain (1975) investigated the essential amino acid content of mycelium of two strains of *Morchella* spp. and found them well with FAD standards except for methionine.

Hatanka (1969) isolated a Cis-3-Amino-1-Proline from fruit bodies of *M. esculenta, M. conica* and *M. crassipes*. It is present in a free state in fruit bodies of *M. esculenta, M. conica,* and *M. crassipes* as well as in their cultured mycelia. He conducted paper chromatographic analysis of over 300 species of higher fungi, and observed that this amino acid was confined to some species of *Morchella* only.

Hatanka (1976) recorded this non-protein amino acid from cultured mycelium of *Morchella esculenta* as well. He investigated that Cis-3-Amino-l-Proline was formed rapidly from L-Proline and accumulated in cultured mycelium of *M. esculenta.* Possible pathway of the formation of this amino-acid was via-3-hydroxy-proline, successively. Moriguchi *et al.* (1979) isolated Cis-3-Amino-1-Proline from fruit bodies of *M. conica, M. crassipes,* and *M. esculenta*. They isolated this from the cultured mycelia of *M. esculenta* also.

Moriguchi *et al.* (1986) partially purified r-glutamyl-trans-peptidase (enzyme I, II, III) from the cell free extracts of the cultured mycelia of *M. esculenta.* The molecular masses of enzymes were 155000(I), 219000(II) and 102000(III). All of them catalyzed both hydrolysis and transpeptidation of various r-glutamyl components. R-L-Glutamyl-Ciz-3-amino-L-proline occurring in the cultured mycelia of this fungus was a good substrate for both reactions.

It is evident that the eight species of *Morchella* (Lakhanpal and Shad, 1989) have never been analyzed simultaneously for nutritive component in India. In the following pages, nutritive analysis of all the eight species from Himachal Himalaya is being presented and the standard methodology is given in Annexure-V.

OBSERVATIONS AND DISCUSSION

Table 7.1 shows that per cent moisture fresh weight of *Morchella* spp. sporophores ranges from 84.15–88.41. Protein content of *Morchella* spp. analyzed ranged from 32.90 to 35.05 g/ 100 gm dry weight (N x 6.25).

Table 7.2 shows that the total free amino acids vary from 6.6 to 11.8 (gm/100 gm) in the sporophores of the *Morchella* species. Sporophores of different species of *Morchella* show total 18 amino acids, namely arginine, alanine, aspartic acid, glycine, cystine, glutamic acid, glutamine, isoleucine, lysine, methionine, phenylalanine, proline, serine, tyrosine, threonine, valine, leucine and tryptophan (Table 7.2).

Out of these eighteen, eleven amino acids were present in the sporophores of *M. semilibera,* twelve in the sporophores of *M. deliciosa, M. esculenta* and *Morchella simlensis* sp. nov. and thirteen in the sporophores of *M, crassipes, M. angusticeps, M. conica* and *M. tibelica.*

Table 7.3 shows that the reducing sugars and non-reducing sugars in different species of *Morchella* vary from 1.3 to 2.8 and 19.5 to 25.5 (g/100 g dry weight), respectively. Starch and carbohydrates range in the sporophores of different species from 12.1 to 19.2 and 21.0 to 27.2 (g/100 g dry weight), respectively.

Table 7.1 Nutritive components of *Morchella* sporophores (g/100 g)

Morchella spp.	*Moisture fresh* ***weight (%)***	*Crude Protein g/100 g Dry weight (N × 6.25)*	*Elemental composition* (g/100 g dry wt.)*						*PPM (Dry weight)*			
			N	**P**	**K**	**Ca**	**Mg**	**S**	**Zn**	**Cu**	**Mn**	**Fe**
M. esculenta	84.15	34.30	5.488	1.25	1.9	0.6	0.17	0.37	154	44.0	52.0	523
M. conica	84.26	32.90	5.264	1.22	1.8	0.5	0.14	0.35	139	39.0	54.0	532
M. angusticeps	88.41	32.90	5.264	1.29	1.9	0.4	0.14	0.34	147	35.0	32.0	527
M. crassipes	86.12	33.80	5.468	1.24	2.0	0.5	0.16	0.21	201	30.0	42.0	529
M. deliciosa	85.10	35.05	5.678	1.26	1.6	0.4	0.15	0.35	169	36.0	48.0	552
M. semilibera	87.37	33.00	5.280	1.20	1.9	0.6	0.15	0.37	140	39.0	51.0	530
M. tibelica	87.40	32.85	5.262	1.28	1.8	0.4	0.13	0.32	148	34.0	54.0	530
M. simlensis sp. nov.	86.20	33.20	5.388	1.24	1.8	0.5	0.18	0.35	157	45.0	53.0	525

Table 7.2 Amino acid composition of *Morchella* spp.

Morchella spp.	*Arginine*	*Alanine*	*Aspartic*	*Cystine*	*Glutamic*	*Glutamin*	*Glycine*	*Isoleucin*	*Lysine*	*Methionine*	*Phenylalanine*	*Proline*	*Seine*	*Tyrosine*	*Threonin*	*Tryptoph*	*Valine*	*Leucine*
M. esculenta	-	+	+	+	+	-	+	+	+	-	-	+	+	+	+	+	-	-
M. conica	-	+	+	-	+	+	+	-	+	+	+	-	+	+	+	+	+	-
M. angusticeps	+	+	-	+	-	-	+	+	+	+	-	+	+	-	+	+	+	+
M. crassipes	-	+	+	+	+	-	+	-	+	+	+	+	+	-	+	+	+	-
M. deliciosa	-	-	+	+	_	-	+	+	+	+	+	+	+	+	-	+	+	-
M. semilibera	-	+	+	-	_	-	+	-	+	-	+	+	+	-	+	+	+	+
M. tibelica	+	+	-	+	-	-	+	+	+	+		+	+	-	+	+	+	+
M. simlensis sp. nov.	-	+	+	+	+	-	+	-	+	+	+	I	I	-	+	+	-	-

+ = Present (Average of triplicate readings)

– = Absent

Table 7.1 shows that the phosphorus, potassium, calcium, magnesium and sulphur range in different *Morchella* species from 1.20 to 1.26, 1.6 to 2.0, 0.4 to 0.6, 0.14 to 0.17, and 0.21 to 0.37 (g/100 g dry wt.), respectively. Zinc, copper, manganese and iron range from 139 to 201, 30.0 to 44.0, 32.0 to 54, and 523 to 552 (ppm dry wt.), respectively.

Nutritive components of some species of *Morchella* from Kashmir valley were earlier analyzed and reported by Kaul (1978) and some species from Himachal Pradesh by Jandaik (1979). A comparison of the components investigated herein and those obtained by Kaul (1978) and Jandaik *et al.* (1979) is presented in the following pages.

Kaul (1978) also recorded 87.38-90.13 per cent moisture content in *Pleurotus* species. Barbero (1969) recorded 80 per cent moisture in fresh specimens of truffles (Tuber spp.), which is higher than the one reported earlier for *Morchella,* but similar to the protein content of other mushrooms. Earlier Singer (1961), Jandaik *et al.* (1979) and Kaul (1981) had reported that the protein content in different species of *Morchella* varies from 20.31-26.87 per cent. Bano *et al.* (1963, 1981) reported 8.9-38.7 per cent protein content in different *Pleurotus* species. Anderson and Fellers (1942), Andriano and Curz (1933), McConnell and Esselen (1947), Wooster (1954), Watt and Merill (1963), Maggioni *et al.* (1968) and Chang (1972) reported 24.9-33 per cent protein content in *Agaricus bisporus.*

In all seventeen amino acids have been recorded in the six species of *Morchella.* They are arginine, alanine, aspartic acid, cystine, glutamic acid, glycine, isoleucine, leucine, methionine, phenylalanine, proline, serine, threonine, tyrosine, valine, tryptophan and lysine.

These results are in accordance with those obtained by Kaul (1978, b) with sporophores of *M. angusticeps, M. crassipes, M. conica* and *V. bohemica* and mycelium of *M. conica* and *M. angusticeps.* Meckellar and Kohrman (1975) had also reported 16 amino acids from the sporophores of *M. crassipes, M, deliciosa, M. esculenta* and *Morchella* species. But they did not report lysine and tryptophan.

The total soluble carbohydrates, reducing sugars, non reducing sugars and starch were found to be 21.0-27.2 per cent, 1.3-2.8 per cent, 19.5-25.5 per cent and 21.1-18.3 per cent on dry weight basis in the carpophores of all the six species of *Morchella* analyzed.

The phosphorus and potash contents were observed to be higher than other mushrooms in the sporophores of the six species of *Morchella* analyzed, supporting observations of Kaul (1978) made for the sporophores of *M. conica, M. esculenta, M. deliciosa, M. angusticeps, and M, hybrida*. However, all the six species of *Morchella* were found to be poor in iron and slightly rich in magnesium. Kaul (1978) and Jandaik *et al.* (1979 a, b) reported similar results. Minerals have also been recorded to be present in sufficient amounts (Kaul 1978).

It is clear from above that these morels possess higher protein content than most of the vegetables. A qualitative amino acid analysis shows that all of them possess lysine and tryptophan, the amino acids which are deficient in vegetable proteins. Methionine is also present in most of the species. Thus, the morel sporophores have better mineral status and can supply some of the rare elements to human body.

Table 7.3 Nutritive components of *Morchella* sporophores* (g/100 g)

Morchella spp.	*Starch*	*Reducing sugars*	*Total soluble (Carbohydrates)*	*Non-reducing sugar*	*Total free amino acids*
M. conica	16.9	1.7	27.2	25.5	7.8
M. deliciosa	12.6	1.5	21.0	19.5	11.8
M. crassipes	19.2	1.3	21.8	20.5	6.6
M. esculenta	12.9	1.6	24.9	23.3	8.2
M. angusticeps	12.1	2.3	26.0	23.6	9.1
M. semilibera	18.3	2.8	24.0	21.2	10.7
M. tibelica	16.8	2.2	26.2	24.8	8.8
M. simlensis sp. nov.	13.8	1.6	24.2	22.8	6.9

*Each figure is average of three readings.

CHAPTER

8 Nutraceutical Potential

CHAPTER OUTLINE

- Introduction and Review of Literature
- Milestones in Nutraceutical Research
- Observations and Discussion

INTRODUCTION AND REVIEW OF LITERATURE

Human beings have been constantly searching for new substances that can improve biological functions and make people healthier, fitter, and increase their longevity. However, various present-day environmental and lifestyle stresses, including pollution, high-pressure work demands, poor dietary habits, nutrient imbalance and advancing age can cause a weakening of human immune system, a lowering of natural body resistance and an increased proneness to disease. Therefore, they are constantly looking for new substances that can steadily and efficiently improve biological functions. A variety of foods and their components are emerging as factors capable of modifying growth, development, performance and disease resistance. Recently, the entire society has turned to plants, herbs, and mushrooms as sources of these enhancers.

Mushrooms have been used as food since time immemorial. Nutritionally, they are a valuable source of healthy food, which is low in calories and rich in carbohydrates, essential amino acids, fibre, important vitamins and minerals. Although the mushrooms have been traditionally exploited for their gastronomic and nutritional appeal, there has always been interest in certain mushrooms for their tonic or medicinal attributes. Mushrooms have long history of medicinal usages in the orient in addition to their nutritional value. An old Chinese and Japanese proverb states that "Medicines and food have common origin" (Northway, 2005). The mushrooms are good example of foods that health-conscious people can enjoy. This interest increased greatly in the 1970s and much research has been done to determine the validity of claims of medicinal properties and the nature of the action of compounds present in these mushrooms (Miles and Chang, 1997). In the last two decades, there has been an upsurge on the use of mushrooms as nutraceuticals and many edible species have been thoroughly investigated and authenticated for medicinal use.

A new term was coined in 1979 to explain all such applications as **"Nutraceutical"** by Stephen DeFelice. Biesalski (2001) defined nutraceuticals as "A food or part of food that provides medical or health benefits, including the prevention and treatment of disease". Nutraceuticals (often referred as phytochemicals or functional foods) are natural, bioactive, chemical compounds that have health promoting, disease preventing or medicinal properties. Nutraceutical is a product produced from foods, but sold in piles, powders and their medicinal forms (Bawa and Khanum, 2003). Lachance (2003) defined nutraceuticals as natural, bioactive, chemical compounds that have health promoting, disease preventing or medicinal properties. They may range from isolated nutrients, herbal products, dietary supplements and diets to genetically engineered 'designer foods' and processed cereals.

Mushrooms have been used medicinally for centuries, particularly in Traditional Chinese and Japanese Medicine (TCMA & JM). In many Asian cultures, mushrooms are used to promote good health and vitality and to increase the body's adaptive capabilities (Lombardi and Rountree, 2002). The early usage of medicinal mushrooms for health benefits was most probably in the form of extracts, health tonics, concentrates, liquors, tinctures, teas, soups and herbal formula, powders and healthful food dishes. These traditional preparations are now termed as 'Mushroom Nutraceuticals' or 'Dietary Supplements' (Smith *et al*., 2002).

The term "**Mushroom Nutraceuticals**" has been coined by Chang and Buswell (1996) to describe those compounds that have considerable potential as dietary supplements and for use in the prevention and treatments of various human diseases without the troublesome side effects that frequently accompany treatments involving synthetic drugs.

A "**Mushroom Nutraceutical**" is a refined/partially defined extractive from either mycelium or fruiting bodies, which is consumed in a capsule or tablet form as a dietary supplement (not as a regular food) with potential therapeutic applications (Dasvila *et al.*, 2003). Therefore, mushrooms have nutritional value plus health benefits. The metabolites from the nutraceuticals may exhibit features such as antitumour, immunomodulating and hypocholesterolemic properties, etc.

Recent researchers have proved beyond doubt that some mushrooms possess pharmacological activities. The potential of medicinal mushrooms is enormous but mostly untapped. There is, therefore, need to screen other mushrooms, wild as well as cultivated, for medicinal attributes. Among the known mushroom varieties, several have been studied for their ability to enhance the human immune system and fight infections. Some well-known medicinal mushrooms are *Ganoderma lucidum* (Ling Zhi, Reishi), *Grifola frondosa* (Maitake), *Lentinus edodes* (Shiitake), *Agaricus blazei* (Hime-matsutake), and *Cordyceps sinensis* (Caterpillar fungus). Some other lesser familiar mushrooms include *Trametes versicolor* (Turkey tail), *Auricularia auricular-judae, Hericium erinaceus* (Lion's mane), *Flammulina velupties, Volvariella volvacea, Pleurotus ostreatus* (Oyster mushroom), *Agrocybe aegerita, Boletus edulis, Schizophyllum commune* and *Tremella* spp., etc. There are many more species of cultivated and wild edible and non-edible mushrooms that have been analyzed for both their nutritional and nutraceutical components but to a lesser extent than those listed above. The active constituents found in mushrooms are polysaccharides, dietary fibres, oligosaccharides, triterpenoids, peptides and proteins, alcohols and phenols, and mineral elements like zinc, copper, iodine, selenium and iron, vitamins, amino acids, etc. These have been found to boost the immune system, have anti-cancerous properties, act as anti-hypercholesterolemic and hepato-protective agents, show anti-HIV activity and anti-viral activity, ameliorate the toxic effect of chemo- and radiotherapy. Many of the species are known to be aphrodisiacs. Many commercial products from these mushrooms are available in the market. Prominent among these are Lentinan from *Lentinus edodes,* Concord Sunchih and Reishi Plus from *Ganoderma lucidum,* Grifon from *Grifola frondosa,* Didanosine from *Cordyceps militaris,* etc. The nutraceuticals have been found to relieve the stress and pressure of modern society by stimulating both basic and secondary immune response of the immune system.

Arising from the awareness of the relationship between diet and disease has evolved the concept of nutraceuticals. In essence, mushroom nutraceuticals are foods that are eaten not only to satisfy functional dietary needs but also elicit additional health needs. There are more than 5,000 species of mushrooms recorded from all over the world, out of which around 2,000 are identified as edible. A large number of the species are yet to be analyzed for their nutraceutical/medicinal potential. Although the nutritional facts and culinary uses of morels are well accepted all over the world, but the medicinal qualities have yet to make into the mainstream. So, the studies were undertaken to record the ethnic information on the medicinal use of *Morchella* and to check the nutraceutical potential of *Morchella* species in the Himalayan region by extracting the nutritional and medicinal bioactive components, like protein contents, minerals, polysaccharides, vitamins, essential amino acids, calorific value, fibre content and cholesterol level.

The stress and pressure of modern society takes a toll on immune health. Those with weakened immunity are more susceptible to infection and disease. The need to maintain or rebuild a healthy defence has led researchers to minerals, plants and fungi in search of natural health-supporting properties. The fungi in particular are showing promise for their ability to enhance immune function (Lombardi and Rountree, 2002). The work on medicinal mushrooms has been consolidated and reviewed by Lombardi and Rountree (2002) and Wasser (2002).

Mushrooms have been devoured as health-food (both as food and medicine) by mankind since ancient times (Zhuang and Mizuno, 1999). Nutritionally, they comprise a valuable source of health food—low in calories and carbohydrates, rich in proteins, essential amino acids, fibres and important vitamins and minerals, including B vitamins, iron, potassium, selenium and zinc (Mattila *et al.*, 2001). Many mushrooms have been used as food and medicine for centuries, especially in the Asian countries where lot of work has been done on medicinal aspects of several edible mushrooms. They are a typical source of physiological active compounds (Halpern, 1999). Most of the research has been cancer related, but studies conducted have shown that mushroom extracts also reduce the blood pressure and cholesterol concentration and enhance the immune system, antiviral and anti-inflammatory properties. They are also used in treatment of anaphylactic shock, and have anti-HIV properties and an increase of oxygen utilization and antioxidant properties (Chang and Miles, 1996).

Approximately 300 mushroom species have been known to have medicinal or nutraceutical properties and another 1800 species have been identified with prospective medicinal properties. *Ganoderma lucidum* (Reishi), *Lentinula edodes* (Shiitake), *Hericium erinaceous* (Lion's mane), *Pleurotus ostreatus* (Oyster mushrooms) and *Grifola frondosa* (Miitake) are just a few of the cultivated mushrooms species that have been analyzed for medicinal value or nutraceutical potential (Zhaung and Mizuno, 1999).

Nutraceuticals may range from isolated nutrients, herbal products, dietary supplements to genetically engineered 'designer' foods and processed products. Doubtlessly, many of these products possess pertinent physiological functions and valuable biological activities (Andlauer and Furst, 2002). These can be grouped into three broad categories: **nutrients, herbals** and **dietary supplements** (Hathcock, 2001).

In Japan, where interest in functional foods has been growing since the early 1980s, these foods have been defined as "Food for specifies Health Use" (FOSHU) in 1991, with the introduction of a branch of the Ministry of Health and Welfare to deal specifically with their regulations (Arai, 1996) and

- Are expected to have a specific effect on health due to relevant constituents of foods.
- Are foods from which allergens have been removed. The effect of such addition or removal has been scientifically evaluated and permission has been granted to make claims regarding the specific beneficial effect on health to be expected from their consumption.
- Should not pose a health or hygiene risk.

At present, different categories of functional ingredients are recognized in Japan as part of FOSHU (Bawa and Khanum, 2003), and the role of each of the functional ingredients has also been identified and assigned as follows: **dietary fibres, polysaccharides, minerals, proteins, vitamins,** etc.

Of the hundreds of known mushroom varieties, several have been studied for their ability to enhance the human immune system and fight infections. Some well-known medicinal mushrooms with benefits for the immune system include reishi (*Ganoderma lucidum*), shiitake (*Lentinus edodes*), maitake (*Grifola frondosa*), and caterpillar fungus (*Cordyceps sinensis*). Some of the less familiar mushrooms include bai mu erh (*Tremella fuciformis*), fu ling (*Poria cocos*), zhu ling (*Grifola umbellata*), lion's mane (*Hericium erinaceus*), *Auricularia aricularejudae* and *Coriolus versicolor* (Chen, 1989; Kawagishi *et al.*, 1990; Chihara, 1993; Kubo and Nanba, 1996; Chang and Buswell, 1999 and Zhuang and Mizuno, 1999). Saini and Atri (1999) explored mushroom diversity for pharmaceutical activity. The various bioactive substances derived from mushrooms are antibacterial, antifungal, anticancerous, antiviral, anti-HIV, hypocholestrolemic and hypolipidemic, anti-diabetic, blood building and immunity rendering, hepatoprotective, anti-coagulant and anti-thrombic effects (Yamura and Cochran, 1976; Chihara, 1992; Bobek *et al.*, 1993; Hobbs, 1995; Saini and Atri, 1999 and Wasser and Weis, 1997, 1999). In fact, mushrooms have an impressive effect on the cardiovascular system and many varieties (maitake, shiitake and *Cordyceps,* etc.) have been reported to reduce total cholesterol levels, reduce the bad cholesterol (low density lipo-proteins) and triglycerides, decrease platelet binding and reduce arterial pressure. Their extracts can affect glycemic levels and inflammatory conditions (Kino *et al.*, 1986; Zhang and Xia, 1990; Zhu *et al.*, 1998 and Nakamura *et al.*, 1999).

In addition to these promising preliminary findings, scientists have noted that mushrooms have definite primary and secondary physiological effects on the human immune system. Wasser and Weis (1999) have reported that cellular components and secondary metabolites of many mushrooms affect the immune system of the host and, therefore, could be used to treat a variety of disease states.

Mushrooms can directly stimulate both basic (lymphocytes, neutrophils, etc.) and secondary immune responses (immunoglobulins Ig E, Ig A, Ig G) of the immune system. This stimulus can increase production of immune defenders such as cytokines and macrophages, which play vital roles in recognizing and removing foreign antigens, as well as releasing chemical mediators including interleukin-1. Some mushrooms have generalized antimicrobial effects, while others have quite specific properties. This dual capability is important because it provides two separate methods of immune enhancement and response, which is important for treating specific microbial infections and disease states such as gram-negative streptococcal and herpetic virus microbial infections, such as sarcoma cancers, leukemia and hepatitis.

MILESTONES IN NUTRACEUTICAL RESEARCH

Only at the end of 1960s did Eastern and Western scientists start to investigate the mechanisms of health effect of mushrooms. The first successful research discovered the antitumour effect of hot water extracts from several mushroom species (Ikekawa *et al.*, 1969). The main active components proved to be polysaccharides, especially α-D-glucans. Chihara *et al.* (1969) isolated from the fruiting bodies of shiitake water-soluble antitumour polysaccharides, which were named "Lentinan" after the generic name of this mushroom. Lentinan demonstrated powerful antitumour activity, preventing chemical and viral tumor development in mice and experimental models (Zakany *et al.*, 1980 a, b).

Substances that have been found to potentiate the immune system include beta-glucans, lentins, polysaccharides, polysaccharides-peptides complexes, triterpenoides, nucleosides and other secondary metabolites (Liu *et al.*, 1988; Borchers *et al.*, 1999 and Ooi and Liu, 2000). Many of these bioactive substances, through their stimulatory effects on the immune system, are showing powerful antitumour, antimutagneic and anticancer activity (Borchers *et al.*, 1999).

Beta-glucan is isolated from shiitake and maitake mushrooms (Mizuno *et al.*, 1995). Beta-glucan binds to macrophages and other phagocytic white blood cells at certain receptors and activates their anti-infection and antitumour activity by stimulating free radical production (Adachi *et al.*, 1994). This, in turn, signals the phagocytic immune cells to engulf and destroy foreign bodies, i.e., bacteria, viruses or tumour cells (Ohno *et al.*, 1996). Beta-glucan has been reported to be safe and well tolerated and could potentially decrease postoperative infections (Babineau *et al.*, 1994 a,b). In Japan, researchers isolated polysaccharides from the mushroom hime-matsutake (*Agaricus blazei* Murill) that proved to have antitumour effects against sarcoma 180 (Mizuno *et al.*, 1990; Fujimiya *et al.*, 1998 and Wang *et al.*, 2001). Antimutagenic effects of the same mushroom have also been authenticated (Delmanto *et al.*, 2001).

All major medicinal mushroom preparations, both cellular compounds and secondary metabolites, are known to have weak antigenicity and no side effects. A very popular and effective preparation was developed from Turkey tail (*Trametes versicolor*) in Japan in 1965. A polysaccharide-peptide from this mushroom (strain CM-101), under the name Krestin (PSK), was developed which exhibits a marked effect against different types of tumours in experimental animals when administered intra-peritoneally or orally. PSK contains 75 per cent glucan and 25 per cent protein. In 1993, Krestin comprised 25% of anticancer drug market in Japan, and sales totalled US$ 350 million (Chang, 1993 and Mizuno, 1999). An analogous product under the name Polysaccharide Peptide (PSP) was developed in China for turkey tail strain Cov-1. The development process for this strain lasted nine years, from 1983 to 1992 (Hiroshi and Takeda, 1993). Mizuno (1993) stated that, in general, a period of 10 years and a total US$ 75 million, or 10 billion yen, is required from the beginning for the development of a new drug to the time it is marketed.

Another α-D-glucans developed and popular in Japan is schizophyllan from split gill. It is especially effective against cervical cancer (Mizuno, 1996). A glucan from mulberry yellow polypore was developed recently in Korea, and an analogous polysaccharide biotechnology from this species has been accomplished in Japan (Mizuno, 2000).

Bawa and Khanum (2003) also categorized minerals as an important nutraceutical source for health enhancement. These include Zinc (Zn), Copper (Cu), Selenium (Se), and Iron (Fe), etc. Zinc is essential for the function of large number of metalloenzymes including carbonic anhydrase (in stomach and RBCs), carboxypeptidases (in pancreas), phosphatases (in bone, blood, and intestine), dehydrogenase and transaminases. It is also essential for the preservation of ribosomal tertiary structure. Cu is essential for the normal development of bone, the central nervous system and connective tissue. Several enzymes contain copper and the activities of these enzymes will be reduced in copper deficiency. These enzymes include sceruloplasmine (ferroxidase) converting Fe^{+2} to Fe^{+3}, cytochrome oxidase, converting H_2 and O_2 into water, etc. Selenium has been shown to be necessary in small amounts for promoting and maintaining good health. It is also essential for the activity of glutathione peroxidase, a principle protection agent against accumulation of H_2O_2 and organic peroxidases within the cells. It has been associated with reduced cancer risk in several epidemiologic studies. Several experimental studies suggest that Se may be one of the most powerful cancer chemopreventive agents in the human diet. As a constituent of Selenoproteins, Se has enzymatic role as antioxidants. Iron deficiency causes anaemia, a major public health problem in developing countries. Iron deficiency can also cause premature births and increased maternal and prenatal mortality.

Reishi, already mentioned as a sacred mushroom in ancient China, has come to occupy a leading place in present-day medicinal mushroom development. The market values of reishi-based healthcare products in 1995 were estimated as US$ 215 million in Taiwan, US$ 3500 million in China, US$ 600 million in Korea, US$ 350 million in Japan (Chang and Buswell, 1999). The physiologically active substances of reishi are water-soluble polysaccharides and alcohol-soluble triterpenoides. Today, 119 different triterpenoides are identified in reishi (Kim and Kim, 1999), about 80 of which are biologically active. The dietary supplements (DS) of Reishi are valued for their immunomodulating, anticancer, antiviral properties. They are used during remission of cancer and by hepatitis–B patients. They also have anti-hyperlipidemic, hypotensive and hypoglycemic actions (Wasser and Weis, 1997).

Some 30 years ago, epidemiologists studying the native population in the Piedade region in the suburbs of San Paulo, Brazil, noted that the rate of occurrence of adult diseases was extremely low, and found an association with the *Agaricus* species, which was a part of the regular diet of the inhabitants of this area (Mizuno, 2002). This mushroom was identified as *Agaricus blazei* Murr., known by common names royal sun *Agaricus,* himematsutake, kawarihaaratake, or almond-flavoured portbello. Experiments conducted in Japan with mice verified that *A. blazei* significantly activates the immune system (Mizuno, 2002). A number of immunity–enhancing, anticancer, and antitumor fractions were isolated from *A. blazei.* The species was shown to be the most effective anticancer mushroom in study comparing its effect with shiitake, maitake (*Grifola frondosa* (Dicks.: Fr.) S.F. Gray, Polyporaceae), reishi, and other medicinal mushrooms. The fraction identified with immune effect include polysaccharides, (1→6)-(1→4)-Beta-D-glucan, polysaccharide-protein complex (ATOM), RNA-protein complexes, and glucomamman (Stamets, 2000; Mizuno, 2002; Mizuno *et al.,* 1990; Ito *et al.,* 1997 and Fujimiya *et al.,* 1998, 2000).

The Japan Cancer Association proved that *A. blazei* is effective against Ehrlich's ascites carcinoma, sigmoid colon cancer, ovarian cancer, breast cancer, lung cancer, and liver cancer, as well as against solid cancers (Mizuno, 2002).

The mushrooms contain a large amount of well-balanced essential amino acids and have a harmonizing effect. Not only polysaccharides and triterpenoides are known as biologically active, a wide ranges of substances from higher Basidiomycetes belonging to different classes of chemical compounds have also been described and their medicinal properties evaluated. These substances represented glyco-lipids schizonellin), compounds derived from the shikimic acid (strobilurins and oudemansins), aromatic phenols (drosophilin, armillasirin, omphalone), fatty acid derivatives (filiboletic acid, podoscyphic acid), polyacetylenes (agrocybin, xerulin), polyketides (caloporoside, hericenones A-H), nucleocodes (clitocene, nebularine), different sesquiterpenes (protoilludanes, marasmanes, hirsutanes, caryophyllanes, etc.), diterpenes (cyathin, striatal), sesterterpenes (aleurodscal), and many other substances of different origin (Mizuno, 1999 and Wasser and Weis, 1999).

Biologically active substances from higher Basidiomycetes possess antifungal, antibacterial, and antiviral properties. They can be used as insecticides and nematocidal agents. In medicine they are used for immunomodulating both humoral and cellular immune factors in the body. Polyfunctional acidic glucuronoxylomannan isolated from jelly mushrooms (*Tremella* spp.), for instance, stimulates vascular endothelial cells, possesses pronounced anti-radiating effects, stimulates hematogensis, demonstrates anti-diabetic, anti-inflammatory, hypocholesterolemic, anti-allergic activities, and shows hepatoprotective effects. It can be recommended to improve immunodeficiency by AIDS, physical stress or ageing, and it prevents senile degeneration of microvessels, maintaining better blood perfusion conditions in vital organs (Wasser *et al.*, 2000).

The use of medicinal mushrooms goes hand in hand with development of their artificial cultivation. The most significant aspect of mushroom cultivation, if managed properly, is to create zero emissions (no waste). Since more than 70 per cent of agricultural and forest materials are non-productive and are wasted in processing, this is a very real advantage (Pope, 2000). Many of these waste materials can be used as substrates to grow mushrooms. This fact gives a basis to the opinion of many researchers in the field that sustainable development of mushrooms and their products in the 21st century can become a **"Non-green Revolution"** (Chang, 1999). The latest research on the nutraceutical potential of mushrooms has opened up new field for mushroom scientists to explore. The medicinal properties of mushrooms also represent a relatively untapped resource for medical applications. There is a need to unearth the vast treasure of wild and cultivated mushrooms for their medicinal properties. Lakhanpal and Rana (2006) have recently given an account of several medicinally useful mushrooms.

Seven species of *Morchella*, viz. *M. angusticeps*, *M. conica*, *M. crassipes*, *M. deliciosa*, *M. esculenta*, *M. tibelica* and *Morchella simlensis.* sp. nov. were analyzed for their nutraceutical potential following standard methods. All the seven species were categorized into two groups. The basis for grouping has been given in the chapter on DNA polymorphism and genetic diversity in *Morchella* spp.

Group A: *M. angusticeps*, *M. conica* and *M. tibelica.*

Group B: *M. crassipes*, *M. deliciosa*, *M. esculenta* and *Morchella simlensis.*

Standard methodologies for analysis of various nutraceutical components are given Annexure–VI. The observations made during the present investigations are as follows:

OBSERVATIONS AND DISCUSSION

Analysis of Nutritional and Nutraceutical Components

a) Proximate composition of moisture, ash, crude fat, dietary fibres and calorific value

Table 8.1 shows the composition of moisture, ash, crude fat, fibre and the calorific value of species of *Morchella* in the two group categories.

Moisture

Moisture content in the species of group A was 84.15%, whereas it was 86.22% in the species of group B. Further, the moisture content in morels collected during rainy season was 90.82 to 92.84%. It is evident that the moisture content is higher in group B than group A and is highest in monsoon morels. This is probably because the relative humidity during rainy season is very high and, therefore, the fruiting bodies are also swollen and larger in size.

Table 8.1 Composition of moisture, ash, crude fibres and calories in *morchella* spp.

	Moisture content (%)	Ash (%)	Crude fat (%)	Crude fibres (%)	Calorific Value (kcal/100 gm)
GROUP–A	84.15	11.27	2.78	16.54	286.42
GROUP–B	86.22	11.17	3.28	18.32	283.00
Monsoon Morels	90.82 - 92.84	-	-	-	-

Values on dry weight basis

Mushrooms in general have high moisture content. Rai and Sohi (1988) recorded the moisture content in mushrooms as 90.1% in *Agaricus bisporus,* 90.2% in *Pleurotus* spp. and 90.1% in *Volvariella volvacea,* respectively. The moisture content of *Morchella* sporophores is in accordance with the findings of Kaul (1977a), Samajpati (1978), Shad (1989) and Sharma (1993). They recorded the moisture content to be in the range of 87.38–90.13% on fresh weight basis in morels. Shad (1989) also credited that the larger size of morels during rainy season is due to the higher moisture content in the fruiting bodies.

Ash content

The ash content in two groups was almost the same being 11.7% in group A and 11.27% in group B (Table 8.1). Similar observations have been made with regard to ash content in other mushrooms. The ash content was observed to be 8-10% in *Agaricus bisporus,* 11.5% in *Volvariella volvacea* and 7.0% in *Lentinus edodes* (Mattila *et al.*, 2002). Sharma (1993) reported the ash content of different species of *Morchella* between 8.0-9.0%, respectively. Therefore, the morels seem to have higher ash content than other mushrooms. Rai (1995) stated that ash content of mushroom is an index of mineral concentration in a species. Hence, the higher concentration of ash in morels gives a clue towards the high concentration of minerals in morels.

Crude fat

Table 8.1 reveals that the crude fat content in group A was 2.78 and 3.28% in group B. Sharma (1993) reported that the crude fat concentration in different species of *Morchella* ranged between 1.6–2.9%, respectively. Huang *et al.* (1985) reported the crude fat content on dry weight basis as 3.1% in *Agaricus bisporus,* 1.6% in *Pleurotus sajor-caju,* 3.0% in *Volvariella volvacea* and 2.1% in *Lentinus edodes*. Mushrooms are low fat food with 2–8% crude fat on dry weight basis (Crisan and Sands, 1978).

Crude fibre

Table 8.1 shows that the crude fibre percentage in group A is 16.54% and 18.32% in group B. Therefore, it was observed that the highest amount of fibre is in group B than group A.

Dietary fibres are declared as nutrient by Nutritional Labelling and Education Act of 1993 (Gordon, 2002). Any food having 2.5 g or greater amount can be claimed as good source of fibre. Cheung (1996, 1997) calculated total dietary fibre (TDF) of *A. bisporus* (white and brown strains), *L. edodes*, and *P. pulmonarina* as 18.2, 36.6 and 33.1%. Sharma (1993) calculated total dietary fibres in *Morchella* species between 15.3–21.2%. Thus, the morel species have a very good concentration of dietary fibres.

Calorific value

The calorific value is 286.42 kcal/100 g in group A and 283.0 kcal/100 g in group B, respectively on dry weight basis (Table 8.1). Sharma (1993) reported that the total calorific value of *Morchella* species varied from 219.18 to 306.20 kcal/100 g.

It is clear from the above account that morels have good amount of ash, crude fiber, calorific value, dietary fibres and lesser amount of fats.

b) Protein composition

The protein content of morels species in group A is 28.3% and 30.59% in group B and it is 29.23% in morels collected during rainy season on dry weight basis. It is clear that the total protein content among groups A and B of *Morchella* species varied from 28.3 to 30.59% on dry weight basis. The protein content of monsoon morels is 29.29% (Table 8.2). It is, therefore, inferred that the monsoon morels have the same nutritive value as that of spring morels and are not nutritionally different or low in protein content as is the notion with people. Earlier Singer (1961), Jandaik *et al.* (1978) and Kaul (1981) reported that the protein content in different species of *Morchella* varies from 20.31–26.87%, but Shad (1989) and Sharma (1993) reported the protein content to be 28.0–31.0%, which is in accordance with the above results. Mattila *et al.* (2002) has investigated the protein content of three commercially cultivated mushrooms viz. *A. bisporus, P. ostreatus* and *L. edodes*. They observed that *A. bisporus* (white and brown strains) has 20.09 and 20.07% net protein, respectively. These values are higher than the values of *P. ostreatus* and *L. edodes* (10.97 and 10.80% net protein level). It is clear from the observations that the protein content in morels and other mushrooms is quite high.

c) Cholesterol level

Table 8.3 reveals that cholesterol is almost absent in the species of morels investigated in groups A and B. It has also been reported earlier that cholesterol is almost non-existent in mushrooms. Among sterols, ergosterol is reported to be most abundant (Hung *et al.*, 1985). Mattila *et al.* (2002) and Beelman and Edwards (1989) also reported only ergosterol, which is present in mushrooms (*A. bisporus, P. ostreatus, P. sajor-caju*, and *L. edodes*) and is the precursor of vitamin D, and cholesterol is totally absent in all of them.

d) Polysaccharide concentration

Table 8.4 shows the composition of polysaccharide in *Morchella* species. Rhamnose

Table 8.2 Protein content in species of *Morchella*

	Protein content (%)
GROUP–A	28.23
GROUP–B	30.59
Monsoon Morels	29. 29

Values on dry weight basis

(oligosaccharide) is 0.016 and 0.041%; dextrose (monosaccharide) is 0.099 and 0.079% in group A and B of *Morchella* species, respectively. Both A and B have similar presence maltose (oligosaccharide) 0.006 mg/100 g, whereas raffinose (oligosaccharide), mannose (monosaccharide) and fructose (monosaccharide) are not within delectable limits.

Table 8.3 Cholesterol level in species of *Morchella*

	Cholesterol level (%)
GROUP–A	-
GROUP–B	-

Values on dry weight basis

The most relevant nutraceutically important components of mushrooms are the polysaccharides. The monosaccharides undergo a condensation reaction to form long chain, either single or with branches to form polysaccharides (Boyle, 2000). They generally act as host-mediated anticancerous drug and are antitumor in function. Mizuno *et al.* (1990) extracted the antitumor active polysaccharides from medicinal mushrooms. These polysaccharides consist of xylose, mannose, arabinose, galactose, fructose, maltose, glucose, raffinose, rhamnose, etc. (Wang *et al.*, 1995). Polysaccharides generally activated macrophages, T-lymphocytes and other immune effector cells that modulate the release of cytokines, which may in turn account for their antitumor and antimicrobial properties (Mizuno, 1995). Wasser (2000) obtained antitumor polysaccharides from the cell wall of fungus.

Table 8.4 Composition of polysaccharides in *Morchella* species

Sr. No.	*Polysaccharides (Sugars)*	*Group A (%)*	*Group B (%)*
1.	**Rhamnose**	0.016	0.41
2.	**Dextrose**	0.099	0.079
3.	**Maltose**	0.006	0.006
4.	**Raffinose**	Below detection range	Below detection range
5.	**Mannose**	Below detection range	Below detection range
6.	**Fructose**	Below detection range	Below detection range

Values on dry weight basis

Different types of polysaccharides have been isolated from different mushrooms but their concentrations have not been revealed. In *Ganoderma lucidum* various polysaccharides viz. G009, Ganoploy, Gl-P and Gl-37, etc. inhibit the ROS (Reactive Oxygen Species), which have been implicated during pathophysiology of cancer, activate macrophages, T-lymphocytes and natural killer cells (Gao, 2000); enhance the immune response of the patients in advance stage of cancer (Cassileth, 2000 and Jacobson *et al.*, 2000); and have antitumour and anti-angiogenic properties (Lee *et al.*, 1995) and immunopotentiating properties (Nakashima *et al.*, 1979). In *Lentinus edodes* polysaccharides (Lentinan, KS-2) have anti-tumour and antiviral activities (Mizuno, 1995), activate macrophages and have anti-HIV activity. The polysaccharides present in *Grifola frondosa* and *Agaricus blazei* have antitumour properties. In *Cordyceps sinensis* polysaccharides (Cordycepic acids) are effective in regulating blood sugar (Kino *et al.*, 1986), and have anti-metastatic effects (Nakamura *et al.*, 1999) and anti-tumor properties (Bok *et al.*, 1999). The polysaccharides derived from *Schizophyllum commune* (Sonifilan, Sizofiran, Schizophyllan and Sizofilan) show cytostatic effects in

Sarcoma 180 tumor xenographs. Schizophyllan has also been shown to increase the overall survival of patients with head and neck cancers (Kimura *et al*., 1994).

Wasser (2000) isolated 1.38% of polysaccharides from sporocarps of *Ganoderma lucidum*. Chang (1999) reported 8-10% polysaccharides in *Flammulina velutipes*. Gunde Cimerman (1999) also analysed the polysaccharide fraction of *Pleurotus* species, which contains pentoses, hexoses, sucrose, and amino sugars, etc. upto 2-4%. All the higher Basidiomycetes contain polysaccharides of various concentration levels (0.1–4.0%), constituting a major nutraceutical bioactive component (Wasser and Weis, 1999).

Thus, while comparing the concentration of polysaccharides in morels and other nutraceutically important mushrooms, the observations revealed that morels also possess good concentrations of polysaccharides, so they have also been considered nutraceutically important.

e) Nucleic acid concentration

Table 8.5 shows the presence of nucleic acid bases in *Morchella* species. The bases uracil and cytosine are present as 0.036 and 0.029% in group A and B of morels, respectively. The bases adenine, thymine and guanosine are present in very small quantities which are below the sensitivity range of HPLC column. So, they were not detectable. Nucleic acids are important nutraceutical components in mushrooms, but very few have been analyzed for their concentrations.

Nucleic acids are also one of the important active constituents in many medicinally important mushrooms. They impart anti-platelet aggregation, antibacterial activity, and antiviral effects. Shimizu *et al*. (1985) and Kawagishi (1995) reported the presence of nucleic acid (adenosine, cytosine, uracil, RNA) in *Ganoderma lucidum*, but in very minor quantities (0.0002–0.003%). These all nucleoside bases prevent thrombosis. Likewise, Hokama and Hokama (1981) reported the presence of nucleic acid bases in *Lentinus edodes* showing significant platelet agglutination inhibitory effects (anti-thrombotic activity). Later on, Chen and Chu (1996) reported the presence of nucleosides in *Cordyceps* spp., viz. cordycepin [3′ deoxyadenosine]. These compounds interrupt the DNA replication and the DNA inhibition is probably responsible for mechanism of antiviral effects (Liu and Zheng, 1993).

In other medicinally important mushrooms like *Grifola frondosa, Coriolus versicolor, Agaricus blazei, A. bisporus, Volvariella volvacea,* and *Auricularia* spp., etc. the nucleic acids have been reported but their concentration has not been revealed. Whatever concentration has

Table 8.5 Composition of nucleic acid bases in *Morchella* species

Sr. No.	*Nucleic Acid (Bases)*	*Group A (%)*	*Group B (%)*
1.	**Cytosine**	0.036	0.036
2.	**Uracil**	0.29	0.29
3.	**Thymine**	Below detection range	Below detection range
4.	**Adenine**	Below detection range	Below detection range
5.	**Guanine**	Below detection range	Below detection range

Values on dry weight basis

been reported in mushrooms like *Ganoderma lucidum, Lentinus edodes,* the morels seem to have better concentration of nucleic acids than other mushrooms and therefore can be considered as nutraceutically important.

f) Mineral contents

Table 8.6 shows the concentration of minerals in group A and B of morels. The mineral profile reveals that in *Morchella* species selenium (Se) content ranged between 2.03 and 4.63 mg/100 g in the dried samples of groups A and B. Likewise Iron (Fe) content as 4.381 mg/ 100 g in group A and 80.18 mg/100 g in group B. Calcium (Ca) is 5.40 mg/100 g in group A and 5.0 mg/100 g in group B. The zinc (Zn) content is 11.14 mg/100 g in group A and 26. 77 mg/100 g in group B, while manganese (Mn) is 2.47 mg/100 g in group A and 3.17 mg/100 g in group B. Copper (Cu) is 2.44 mg/100 g in group A and 3.18 mg/100 g in group B. Sodium (Na) and potassium (K) are 81.94 and 404.57 mg/100 g in group B and 36.91 and 1668.95 mg/100 g in group A. It is evident from Table 8.6 that the concentration of Se, Fe, Zn, Mn, Cu, Na and K is higher in group B while Cu is present in higher concentration in group A.

Bano *et al.* (1981) and Bisaria *et al.* (1987) assessed the mineral content in *Pleurotus* species. They reported the following concentrations: Zn (2.6-2.9 ppm); Cu (12.2 ppm); Na (60–78 ppm), and Ca (20-40 ppm). In comparison, the species of *Morchella* possess much higher mineral concentrations: Zn (11.14–26.77 mg/100 g); Cu (2.44–3.18 mg/100 g); Na (36.91–81.99 mg/100 g); Ca (5.0–5.40 mg/100 g), respectively. They reported the K content in *Pleurotus* spp. between 2240–3260 mg/100 g d.w. and Spaulding and Beelman (2003) reported the K content of *A. bisporus* to be 3500 mg. In comparison, the morel species possess higher content of K (1668.95 - 4041.57 mg/100 g).

Allen (1997) highlighted some of the nutraceutically important elements in medicinally important mushrooms like *G. lucidum, L. edodes, A. blazei* and *G. frondosa,* but he did not mention the concentration at which these elements were important. He reported that Cu is essential for building bones and teeth, maintaining bone strength, for nerve muscle and glandular functions and for proper functioning of the thyroid; Fe helps in energy production,

Table 8.6 Composition of minerals in *Morchella* species

Sr. No.	*Minerals*	*Group A*	*Group B (%)*
1.	**Selenium (Se)**	2.03 µg/100 g	4.63 µg/100 g
2.	**Iron (Fe)**	43.87 mg/100 g	80.18 mg/100 g
3.	**Zinc (Zn)**	11.14 mg/100 g	26.77 mg/100 g
4.	**Manganese (Mn)**	2.47 mg/100 g	3.17 mg/100 g
5.	**Copper (Cu)**	2.44 mg/100 g	3.18 mg/100 g
6.	**Sodium (Na)**	36.91 mg/100 g	81.94 mg/100 g
7.	**Potassium (K)**	1668.95 mg/100 g	4041.57 mg/100 g
8.	**Calcium (Ca)**	5.40 mg/100 g	5.0 mg/100 g

Values on dry weight basis

and helps to carry and transfer oxygen to tissues; Mn is essential for healthy nerve and muscle function, bone formation and may help to prevent premenstrual syndrome (PMS); Se act as an antioxidant, essential for healthy functioning of the heart muscle and Zn is essential for cell reproduction, normal growth and development in children, wound healing, production of sperm and testosterone.

Mateljan (2005) assessed the mineral content in *Agaricus bisporus*. He reported Se to be 36.85 µg/100 g, Cu to be 0.71 mg/100 g, Mg to be 0.20 mg/100 g, Fe to be 0.57 mg/100 g and Zn to be 1.56 mg/100 g. Se is required for the proper functioning of the antioxidant system. The antioxidant system works to reduce the levels of damaging free radicals in the body. Selenium is an important co-factor for body's most important internally produced antioxidant, glutathione peroxidase. Copper is helpful in reducing the symptoms of rheumatism and is also necessary for the activity of lysyl oxidase, an enzyme used for the flexibility in blood vessels, joints and bones. Magnesium is a co-factor of a key oxidative enzyme called superoxide dismutase. Iron is a part of haemoglobin. Zinc affects the fundamental processes like immune system, wound healing, normal cell division, and stabilizes blood sugar levels and body's metabolic rates, necessary for the optimal sense of taste and smell and prevents illness.

Stingve (1977) observed variable concentration of Se in different mushrooms. He reported it to vary between 0.012–20.0 µg/100 g d.w. He observed the highest concentration of Se in *Boletus edulis*. Mattila *et al.* (2002 a) pointed out that Se content in *A. bisporus* (3.2–1.4 mg/kg dry weight for brown and white strains, respectively) is greater than that of *P. ostreatus* and *L. edodes* (0.15 & 0.04 mg/kg d.w., respectively).

Spaulding and Beelman (2003), however, reported the Se concentration in brown and white strains of *A. bisporus* grown at commercial farms to be between 1.0–2.1 µg/100 g d.w. At this concentration, Se functions as a necessary cofactor in the enzyme system of glutathione peroxidase that protects cell membrane lipids from damaging effects of free radicals (Reilly, 1996). Selenium is also known to decrease the incidences of cancer (Clark *et al.*, 1996). Recent studies show that Se-enriched *Agaricus bisporus* strains significantly reduced the occurrence of DMDA (7, 12 dimethylbenz (A) anthracene)–induced mammary epithelial cell DNA adducts (Spolar *et al.*, 1999).

Selenium is a mineral micronutrient that has generated much interest in nutraceutical and medical research and the food industry. It has numerous physiological functions but is best known as a necessary co-factor for the glutathione peroxidase enzyme system. This system is responsible for removing free radicals from the body, thus reducing oxidative damage. Owing to this potent antioxidant effect, selenium has been associated with the reduction of various chronic, degenerative diseases, including several types of cancers. *Agaricus* mushrooms normally contain 1-2 ppm (d.w.) Se, which makes them very good sources for removing free radicals (Werner and Beelman, 2001).

Therefore, it is evident that *Morchella* species have higher content of Se (2.03-4.63 µg/100 g d.w.) than cultivated mushrooms (especially *Agaricus bisporus*) and are, therefore, nutraceutically important and can be used for the removal of various free radicals from the body.

Thus, it is clear that *Morchella* species are nutraceutically important and richer in mineral contents (Se, K, Zn, Cu, Na, and Ca) in comparison to other edible mushrooms like *Agaricus bisporus, Ganoderma lucidum,* and *Lentinus edodes*, etc. and therefore can be further investigated and exploited for their medicinal attributes.

g) Vitamin concentration

Table 8.7 shows the quantity of water-and fat-soluble vitamins on dry weight basis in both A and B groups of *Morchella* species.

Water-soluble vitamins

The amount of vitamin C (Ascorbic acid) is 3.49 mg/100 g in group A, while it is 3.59 mg/100 g in group B. Vitamin B_3 (Niacin) is 8.62 mg/100 g in group A, while it is 8.98 mg/100 g in group B. Vitamin B_6 (Pyridoxine) was observed to be below the detection range of HPLC in both the groups, therefore not detected. Vitamin B_1 (Thiamine) is 0.15 mg/100 g in group A and it is 0.264 mg/100 g in group B, while vitamin B_2 (Riboflavin) is 0.21 mg/100 g in group A and 0.252 mg/100 g in group B. These observations reveal that there is higher concentration of water-soluble vitamins in group B than group A.

Fat-soluble vitamins

Vitamins A (Retinal) and D (Ergocalciferol) were in the detection limits but vitamins E (Tocopherol) and K were below the sensitivity range of HPLC Column, hence not detectable. The content of vitamins A and D is 0.0003 and 0.00135 mg/100 g in group A and 0.0004 and 0.0014 mg/100 g in group B.

According to Bawa and Khanum (2003), vitamins are required in small amounts in foods

Table 8.7a Composition of water-soluble vitamins in *Morchella* species

Sr. No.	*Water-soluble Vitamins*	*Group A (mg/100g)*	*Group B (%) (mg/100g)*
1.	**Vitamin C (Ascorbic acid)**	3.49	3.59
2.	**Vitamin B_3 (Niacin)**	8.62	8.98
3.	**Vitamin B_6 (Pyridoxine)**	Below detection range	Below detection range
4.	**Vitamin B_1 (Thiamin)**	0.15	0.264
5.	**Vitamin B_2 (Riboflavin)**	0.21	0.252

Values on dry weight basis

and cannot be synthesized by the body. They are used for the maintenance of optimal health, growth and reproduction. Many vitamins act as coenzymes whereas others have no single role but perform certain essential functions. Vitamins and minerals are although basic nutrients of the body, but are also exploited as nutraceuticals for therapeutic purpose. Vitamin B_1 helps in energy metabolism, correct functioning of heart and nervous system. Vitamin B_2 is involved in carbohydrate, polysaccharide and fat metabolism and helps to release energy into cells. Niacin (Vitamin B_3) is involved in energy producing reactions in cells and aids the functioning of nervous system of the body. Vitamin B_6 is essential for proper utilization of proteins. Vitamin C helps to maintain bone, teeth and blood vessels, is important in formation of collagen for body structure, and is involved in disease prevention.

Table 8.7b Composition of fat-soluble vitamins in *Morchella* species

Sr. No.	Fat-Soluble Vitamins	Group A (mg/100 g)	Group B (mg/100 g)
1.	**Vitamin A (Retinal)**	0.0003	0.0004
2.	**Vitamin D (Ergocalciferol)**	0.00135	0.0014
3.	**Vitamin E (Tocopherol)**	Below detection range	Below detection range

Values on dry weight basis

Vitamin A helps in normal growth, healthy skin, eyes, teeth, reproduction and disease prevention. Vitamin D is necessary for strong teeth and bones and regulates calcium/ phosphate metabolism.

Mateljan (2005) reported the vitamin contents in *Agaricus bisporus* as follows: vitamin B_I 0.13 mg/100 g, vitamin B_2 - 0.69 mg/100 g; vitamin B_3 - 5.39 mg/100 g; vitamin B_6 - 0.16 mg/100 g; vitamin C - 3.0 mg/100 g; vitamin A - 0.0001 - 0.0002 mg/100 g; vitamin D - 0.0002 mg/100 g. He also highlighted the role of vitamins like Bawa and Khahum (2003) as follows: vitamin B_1 helps to convert food into energy and is essential in neurological functions; vitamin B_2 helps in energy production and other chemical processes in the body, maintenance of healthy eyes, skin and nerve function; vitamin B_3 helps to produce the generic materials of cells, formation of red blood cells, maintenance of central nervous system and synthesize amino acids and is involved in the metabolism of fats, protein and carbohydrates. It is also helpful in reducing cholesterol levels and in preventing osteoarthritis; vitamin B_6 is needed to convert homocysteine, a dangerous molecule that can directly damage blood vessel walls, into other substances. Vitamin C acts as an antioxidant necessary for healthy bones, gums, teeth and skin, helps in wound healing, may prevent common cold and attenuate its symptoms. Vitamin A is an antioxidant, essential for growth and development. It maintains healthy vision, skin and mucous membrane and may aid in the prevention and treatment of certain cancers and in the treatment of certain skin disorders. Vitamin D is essential for formation of bones and teeth and also helps the body to absorb and use calcium, but is in very lesser concentration in this mushroom.

Mushrooms have been reported to be a good source of many vitamins, especially those of water-soluble group (B - complex and C), but are relatively very poor in fat-soluble vitamins - A, D, E, K (Crisan and Sands, 1978 and Chang and Miles, 1989).

Thiamine (B_1) content in various mushrooms, i.e. in *V. volvacea* has been reported to be 0.35 mg; in *A. bisporus* 1.14 mg; in *Pleurotus* species 1.16 mg and 7.8 mg in *L. edodes.* But the content of thiamine in morels is 0.15-0.264 mg, which is higher than *V. volvacea* but lesser than others.

The niacin (B_3) has been reported to be 54.9 mg in *L. edodes*, 55.7 mg in *A. bisporus*, 64.88 mg in *V. volvacea* and 46.0-1.8.7 mg in *Pleurotus* spp. (Chang and Miles, 1989). In the species of Morchella, the niacin content is only 8.62–8.98 mg, which is less than that reported in other cultivated mushrooms.

A. bisporus mushroom is also known to be a good dietary source of B-complex vitamins especially vitamin B2 (riboflavin), i.e. 1.7 mg/100 g (USDA, 2001). Mattila *et al.* (2002 a)

reported that *A. bisporus* white and brown strains contained 0.39 and 0.33 mg/10 g riboflavin, while *P. ostreatus* and *L. edodes* contain only 0.20 and 0.15 mg/100 g, respectively. But as is apparent in morels, the concentration is 0.21-0.25 mg/100 g, which is higher than that reported for other mushrooms.

Mushrooms also contain vitamin C (ascorbic acid) Crisan and Sands, 1978; Bano and Rajarathnam, 1982). *A. bisporus* contains about 3.0 mg, *P. ostreatus* has 2.0-3.0 mg and *Pleurotus sajor-caju* has 4.0-5.0 mg (Beelman and Edwards, 1989). *Morchella* spp. have 3.4-3.59 mg of vitamin C, which is higher than that reported in other mushrooms.

It is clear from the above account of vitamins that from amongst the water-soluble vitamins, the morels have higher concentration than other edible mushrooms in terms of vitamins B_1, B_2 and C, while the concentration of B_3 is less than the others.

Mattila *et al.* (2002 a) observed that vitamin D is almost totally absent in cultivated mushrooms, since they thrive in the absence of sunlight. But in the species of *Morcheella* investigated, a good amount of vitamin D (0.00135 – 0.014 mg/100 g) is present. This is probably because morels are collected from wild and receive enough light. Similar is in the case of vitamin A; Mateljan (2005) reported 0.0001-0.0002 mg/100 g in *Agaricus bisporus* fruiting bodies, while *Morchella* spp. have higher concentration of vitamin A (0.0003-0.0004 mg/100 g).

Therefore, as compared to other edible mushrooms, morels possess a good amount of water-soluble (B_1, B_2 and C) as well as fat-soluble vitamins (A, D, K).

h) Amino acid concentration

Table 8.8 shows the presence of essential amino acids viz. histidine, threonine, methionine, tyrosine, phenylalanine, valine, isoleucine and leucine in morels. The amount of histidine is 3.50 mg/100 g in group B and 2.44 mg/100 g in group A. The amount of threonine and methionine is same in both groups A and B, i.e. 2.4 mg/100 g and 2.45 mg/100 g. The amount of tyrosine is 2.55 mg/100 g in group A and 2.32 mg/100 g in group B. Phenylalanine is 2.34 mg/100 g in group A and 2.30 mg/100 g in group B. Similarly, lysine, isoleucine and leucine are 2.45, 2.27 and 2.16 mg/100 g in group A and 1.94, 1.9, 1.66 mg/ 100 g in group B. It is, therefore, clear that the amino acids viz. tyrosine, phenylalanine, valine, lysine, leucine are present in higher concentrations in group A than group B, while

Table 8.8 Essential amino acid content of *Morchella* species (g/100 g)

Sr. No.	*Amino Acids*	*Group A*	*Group B*
1.	**Histidine**	2.44	3.50
2.	**Threonine**	2.49	2.49
3.	**Methionine**	2.45	2.45
4.	**Tryptophan**	2.55	2.32
5.	**Phenylalanine**	2.34	2.30
6.	**Valine**	2.38	2.34
7.	**L-Lysine**	2.45	1.94
8.	**Isoleucine**	2.27	1.99
9.	**Leucine**	2.16	1.66

Values on dry weight basis

only histidine has greater concentration in group B in comparison to group A. Whereas, threonine and methionine have similar concentrations in both the groups A and B, respectively.

Amino acids are nutraceutically very important. They are anti-hypersensitive, anti-immune diabetic, anti-hepatitis B and immunomodulatory in function (Kino *et al.*, 1989). The mushrooms that have been exploited and investigated for their amino acid content to a greater or lesser extent are *Ganoderma lucidum, Lentinus edodes, Grifola frondosa, Agaricus blazei, A. bisporus, Cordyceps sinensis, Pleurotus* spp., *Tremella* sp., *Flammulina velutipes, Hericium erinaceous, Auricularia* sp., *Schizophyllum commune* and *Trametes versicolor,* etc., but their concentrations have not been revealed and only their functions are mentioned.

L. edodes contains eight essential amino acids and is rich in lysine and leucine. They help in lowering blood serum cholesterol (BSC) and lipid concentration (Yamura and Cochran, 1976). These amino acids also accelerate the excretion of ingested cholesterol and its metabolic deposits, and in humans lower the cholesterol levels (Susuki and Oshima, 1974). Another amino acid derivative, thioproline, blocks the formation of carcinogenic N-nitroso compounds, and Lenthionine, a sulphur containing peptide, has antibacterial and antifungal activity (Hatavani, 2001). *Cordyceps* spp. also contains all essential amino acids, which produce some potent antibiotics, are immune stimulants and antitumor agents (Hobbs, 1986), and are also used for the treatment of respiratory diseases, renal dysfunction, hyperlipidemia and hyperglycemia (Zhu *et al.*, 1998).

Shad (1989) qualitatively determined total free amino acid content in *Morchella* species. He reported that the total amino acid concentration ranges from 6.6-11.8 g/100 g for all the eighteen amino acids present. Later on Sharma (1993) also showed the qualitative estimation of seven essential and two non-essential amino acids in *Morchella* species. Mattila *et al.* (2002) calculated the concentration of all the essential amino acids (cystine, methionine, threonine, isoleucine, leucine, lysine, tyrosine and phenylalanine) in *A. bisporus* (brown and white strains), *P. ostreatus* and *L. edodes* and reported to vary from 23-292 mg/100 g. A comparison of these results with the result of morels shows that morels have higher concentration of amino acids than the cultivated mushrooms.

The nutraceutical potential of mushrooms depends upon the nutritive and medicinal composition. The present investigations on nutraceutical potential of morel fruiting bodies have shown that the *Morchella* spp. possess good nutritional and medicinal attributes. The observations made demonstrate that nutraceutical attributes of *Morchella* species compare favourably with other commercially accepted mushrooms (*Agaricus bisporus, Lentinus edodes, Pleurotus ostreatus, P. sajor-caju,* etc.). *Morchella* species are a better source of polysaccharides, crude fibres, nucleic acids, minerals especially (Se, K, Zn, Cu, Na, and Ca, etc.), vitamins (B_1, B_2, C, A, D and K), proteins, and all essential amino acids. The morels are almost free of cholesterol. All these properties of morels prove them to be of good nutraceutical use.

CHAPTER

9

Rhizomorphosphere Relationships

CHAPTER OUTLINE

INTRODUCTION AND REVIEW OF LITERATURE

Soil is the repository of all life and the laboratory within which are carried out most of the changes that enable life to continue (Gilman, 1967). The soil fungi are invariably components of the microbiota of normal soils. The microorganisms such as bacteria, fungi and actinomycetes of thermophilic, thermotolerant and mesophilic nature play role in various chemical reactions to make the substrate suitable for the growth of any plant (Belskonig *et al.*, 1962; Hayes, 1969 and Stanek, 1972).

The aerial and underground parts of the plants harbour varied populations of microorganisms. The portion of soil surrounding the roots is termed as "Rhizosphere" (Hiltner, 1904) and the root surface "Rhizoplane" (Foster. 1986). Generally, the microbes come on the root surface from the soil. Foster and Marks (1967) recognized non-mycorrhizal roots as having rhizosphere and mycorrhizal roots as having mycorrhizosphere. The rhizosphere is reported to be rich in organic metabolites. The root exudates liberated into it by the tree root, stimulate or inhibit mycelial growth, germination of microsclerotia, active zoospore activity and virulence to various pathogens. The rhizosphere is actually a complex habitat characterized by the presence of many exudates and various nutrients which have been solubilized by the weathering action of the microflora (Davey, 1969).

The term "**Rhizomorphosphere**" is coined to designate the portions of the soil surrounding the rhizomorphs of morel fruiting bodies.

Since rhizosphere microorganisms greatly influence the root activity, it was desired to know whether similar activity and effect is exercised by microorganisms on the rhizomorphs of *Morchella* species or not. To substantiate this, the mycoflora associated with rhizomorphs of *Morchella* species was studied. It was also desired to know whether microflora of morel bearing and non-morel bearing is similar or different. Hence, microorganisms were also isolated from morel bearing and non-morel bearing soils. The standard methodology for isolation of fungi and isolation of pure cultures is given in Annexure–VII.

OBSERVATIONS AND DISCUSSION

The results obtained indicated that there was no difference in the microbial populations of morel bearing and non-morel bearing soils. In all forty-eight species of soil fungi were isolated from both types of soils collected from different forest types (Table 9.1). Twenty-two of these belong to Deuteromycotina, sixteen to Ascomycotina, ten to Zygomycotina and some represent sterile mycelial forms with clamp connections, belonging to Basidiomycotina. No such studies have ever been done earlier with fungi. Rhizosphere and rhizoplane studies have been done mostly with plant root system, so comparison with other researchers on fungi especially morels is not possible.

The details of fungal associations with morels are as follows:

The soil bearing and not bearing *M. angusticeps* was collected from forests composed of *Pinus wallichiana, Cedrus deodara, Abies pindrow-Picea smithiana* and *Quercus semicarpifolia* trees. Twenty-one species of fungi were isolated from these forests (Fig. 9.1). Six were members of Deuteromycotina, seven of Ascomycotina, seven of Zygomycotina and one white sterile

Table 9.1 Rhizomorphosphere fungi in different *Morchella* species

Sr. No.	Name of Fungi Isolated	M. angusticeps	M. conica	M. crassipes	M. deliciosa	M. esculenta	M. semilibera
1.	***Alternaria alternate*** **(Fr.) Kiessler**	+	+	+	+	+	-
2.	***Alternaria*** **sp.**	-	-	-	-	-	+
3.	***Aspergillus flavus*** **Link.**	+	+	+	+	+	-
4.	***A. fumigatus*** **Fresenius**	+	+	+	+	+	-
5.	***A. humicola*** **Chaudhari**	-	+	-	+	-	-
6.	***A. luchuensis*** **Inui**	+	-	+	+	+	+
7.	***A. niger*** **Van Tieghem**	-	+	+	-	-	-
8.	***Cephalosporium acremonium*** **Corda**	-	-	+	-	-	-
9.	***C. aspermum*** **Marchal**	-	-	-	+	-	-
10.	***C. roseo-griseum*** **Saksena**	+	+	-	+	+	-
11.	***Cephalosporium*** **sp.**	+	-	+	+	-	-
12.	***Cladosporium cladosporioides*** **(Pers.) de Vries**	-	-	-	-	+	-
13.	***Cladosporium*** **sp.**	-	-	-	-	+	-
14.	***Fusarium moniliforme*** **Sheldon**	-	+	+	-	-	-
15.	***Fusarium oxysporium*** **Schl. ex Fr.**	-	+	-	-	-	-
16.	***Fusarium*** **sp.**	-	-	+	+	+	-
17.	***Mucor globosus*** **Fisher**	+	-	-	-	+	-
18.	***M. hemimalis*** **Wehmer**	+	-	-	-	-	-
19.	***M. microsporus*** **Namyslowski**	+	-	-	+	-	-
20.	***Mucor*** **sp.**	-	-	+	-	-	-
21.	***Penicillium biforme*** **Thom**	-	-	+	+	+	-
22.	***P. chrysogenum*** **Thom**	+	+	+	+	+	+
23.	***P. citrinium*** **Thom**	+	+	+	+	+	+
24.	***P. funicolosum*** **Thom**	-	+	-	-	-	-
25.	***P. helicum*** **Raper and Fennell**	-	+	-	-	+	-
26.	***P. javanicum*** **Van Beyma**	+	+	+	+	+	-
27.	***P. oxalicum*** **Thom**	-	-	-	-	-	-
28.	***P. paxilli*** **Bainier**	+	-	-	-	-	-
29.	***P. roseum*** **Link**	-	-	-	-	+	-
30.	***Penicillium stoloniferum*** **Thom**	-	+	-	+	-	-
31.	***Penicillium*** **species**	-	-	+	-	+	-
32.	***Phoma hybernica*** **Crimes, O' Connor and Cumminis**	-	-	+	-	-	-
33.	***Rhizopus nigricans*** **Ehrenb**	+	+	+	+	+	-
34.	***R. nodosus*** **Nam Slowski**	-	+	-	-	-	-
35.	***R. oryzae*** **Went and Gerlings**	+	+	-	-	-	-
36.	***R. stlonifer*** **Ehrenb (Fr.) Lind**	-	+	+	-	-	-
37.	***Rhizopus*** **sp.**	+	-	-	+	-	-

(Table Contd.)

(Table Contd.)

38. ***Spicaria elegans*** **Corda**	-	+	-	-	-	-
39. ***Spicaria fisipora*** **Saksena**	-	+	-	-	-	-
40. ***Spicaria griceola*** **Saccardo**	-	+	-	-	-	-
41. ***Trichoderma glaucum*** **Abbott**	-	+	+	-	+	-
42. ***Trichoderma harzianum*** **Rifai**	+	+	+	+	+	-
43. ***Trichodernma roseum***	-	-	-	-	+	-
44. ***Trichoderma viride*** **Pers. ex Fr.**	+	+	+	+	+	+
45. ***Vertlicillium alboatrum*** **Reinke and Borthold**	-	+	-	-	-	-
46. ***Verticillium puniceum*** **Cooke and Ellis**	-	-	-	-	-	-
47. ***Verticillium*** **sp.**	+	+	+	-	+	-
48. ***Zygorhynchus*** **sp.**	+	-	-	-	-	-
49. **White sterile mycelium**	+	+	+	+	+	+

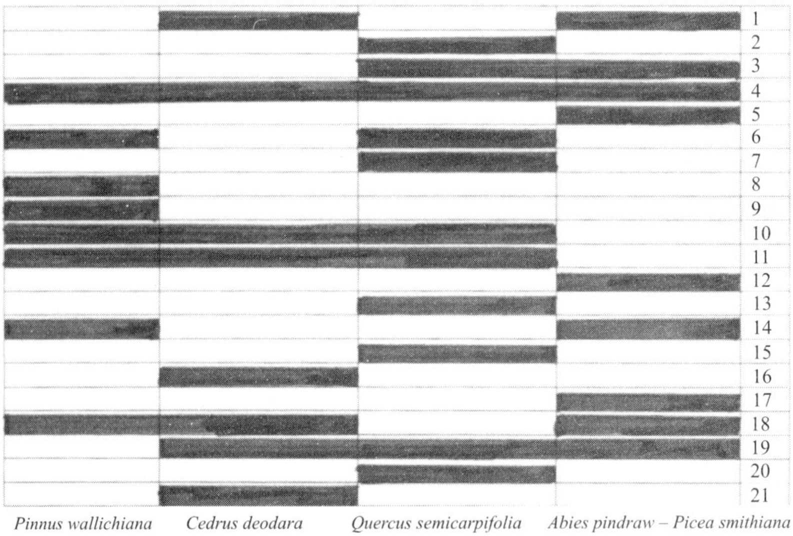

1. White sterile mycelium, 2. *Zygorhynchus* sp., 3. *Verticillium* sp., 4. *Trichoderma viride*, 5. *T. harzianum*, 6. *Rhizopus* sp., 7. *R. oryzae*, 8. *R. nigricans*, 9. *Penicillium paxillae*, 10. *P. javanicum*, 11. *P. citrinum*, 12. *P. chrysogenum*, 13. *Mucor microsporus*, 14. *M. heimalis*, 15. *M. globosus*, 16. *Cladosporium* sp., 17. *Cephalosporium roseum-griseum*, 18. *Aspergillus luchuensis*, 19. *A. fumigatus*, 20. *A. flavus*, 21. *Alternaria alternata.*

Fig. 9.1 Fungi associated with *M. angusticeps* bearing soils in different forest types

mycelial form. From *inus wallichiana* forest soil were isolated *javanicm**, *citrinum, T viride, R paxillae, R nigricans, Rhizopus* sp., *A luchuensis* and *Mucor heimalis* From *Cedrus deodara* forest soil *A alternata javanicum, T viride, A luchuensis, A fumigatus, ladosporium* sp. and white sterile mycelium were isolated. In *Quercus* forest soil, *T viride, A flavus, A*

*The authorities for the species have been given in Table 9.1.

fumigatus, P. javanicum, R. oryzae, Verticillium sp., *M. globosum, M. microsporus, P. citrinum, Rhizopus* sp. and *Zygorhynchus* sp. were isolated. From the forest soil of *Abies Pindrow- Picea smithiana* were isolated, *T. viride, T. harzianum, P. chrysogenum, A. fumigatus, A. luchuensis, Cephalosporium roseum-griseum, Verticillium* sp., *Mucor heimalis* and white sterile mycelium.

It is clear that in *M. angusticeps,* out of total 21 fungi isolated, *Trichoderma viride* was present in the soils of all the four forest types, and *Aspergillus fumigatus, Aspergillus luchuensis, P. citrinum, P. javanicum* were present in soils of three forest types. *Mucor heimalis, Verticillium* sp. and white sterile mycelium were present in the soils of two forest types. *Alternaria alternata, Aspergillus flavus, Cephalosporium roseum-griseum, Cladosporium* sp., *Mucor globosus, M. microsporus, Penicillium chrysogenum, P. paxillae, Rhizopus nigricans, R. oryzae, Rhizopus* sp., *Trichoderma harzianum,* and *Zygorhynchus* sp. were present in the soils of one forest only.

The soil bearing *M. conica* was collected from the forests consisting mainly of *Pinus wallichiana, Cedrus deodara, Abies pindrow-Picea smithiana* and *Quercus semicarpifolia* trees. Twenty-seven species of fungi were isolated from the soils of these forests (Fig. 9.2). Out of these twelve species belong to Deuteromycotina, nine to Ascomycotina, five to Zygomycotina and one was white sterile mycelial form. The soil from *Pinus wallichiana* yielded *Penicillium javanicum, Trichoderma viride, P. citrinum, Rhizopus nigricans, R. stolonifer, R. oryzae, Cephalosporium roseum-griseum, Aspergillus humicola, A. niger, Spicaria elegans, Verticillium*

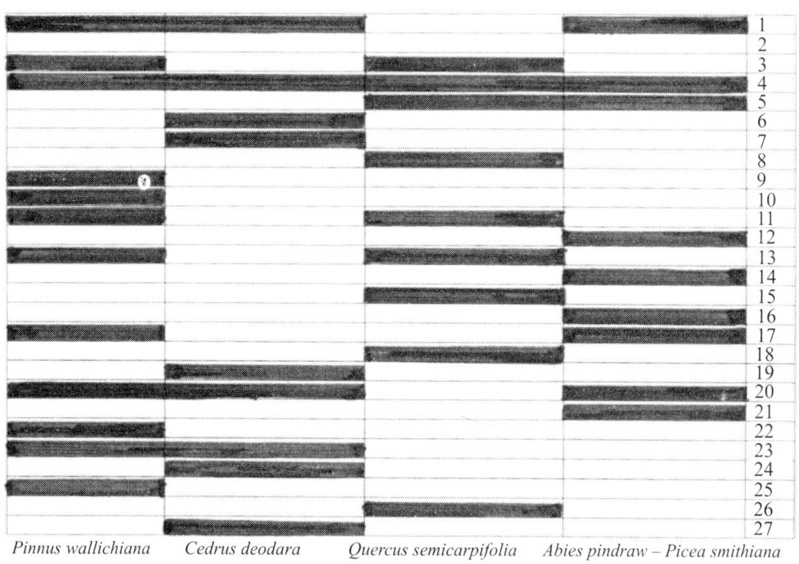

1. White sterile mycelium, 2. *Verticillium* sp., 3. *Verticillium albo-atrum,* 4. *Trichoderma viride,* 5. *T. harzianum,* 6. *T. glaucum,* 7. *Spicaria gracea,* 8. *S. fusispora,* 9. *S. elegans,* 10. *Rhizopus stolonifer,* 11. *R. oryzae,* 12. *R. nodosus,* 13. *R. nigricans,* 14. *Fusarium oxysporium,* 15. *F. moniliforme,* 16. *Penicillium stoloniferum,* 17. *P. javanicum,* 18. *P. helicum,* 19. *p. funiculosum,* 20. *P. citrinum,* 21. *P. chrysogenum,* 22. *Cephalosporium roseum-griseum,* 23. *Aspergillus humicola,* 24. *A. fumigatus,* 25. *A. niger,* 26. *A. flavus,* 27. *Alternaria alternata.*

Fig. 9.2 Fungi associated with *M. conica* bearing soils in different forest types *alboatrum* and white sterile mycelium. From *Cedrus deodara* forest soil, *Alternaria alternata, T. viride, T. glaucum, P. citrinum, Aspergillus humicola, A. fumigatus, P. funiculosum, Spicaria gracea* and white sterile mycelium were isolated. In *Quercus semicarpifolia* forest soil, *T. viride, T. harzianum, A. flavus, R. oryzae, R. nigricans, Fusarium moniliforme, P. flavus, R. oryzae, R. nigricans, Fusarium monoliforme, P. helicum, Spicaria fusiporus,* and *Verticillium moniliforme* and *Verticillium alboatrum* were present, and from the forest soil of *Abies pindrow - Picea smithiana,* the fungi isolated were *T. viride, T. harzianum, P. chrysogenum, P. citrinum, P. javanicum, Fusarium oxysporum, R. nodosus* and a white sterile mycelium.

It is clear that in *M. conica* out of the 26 fungi isolated, *T. viride* was present in the soils of all four forest types and *P. citrinum* and white sterile mycelium in the soils of three forest types. *A. humicola, P. javanicum, R. nigricans, R. oryzae, T. harzianum* and *Verticillium alboatrum* were present in the soils of two forest types and *Alternaria alternata, Aspergillus flavus, A. niger, Cephalosporium-roseum-griseum, Fusarium moniliforme, F. oxysporum, Penicillium chrysogenum, P. helicum, P. funiculosum, P. stoloniferum, R. nodosus, R. stolonifer, Spicaria elegans, S. fusiporus, S. griceola, T. glaucum, and Verticillium* species were present only in soils of one forest type.

The soil bearing *M. crassipes* was collected from forests composed of *P. wallichiana, Cedrus deodara, P. smithiana-Abies pindrow* and *Quercus semicarpifolia* trees. In all twenty-three fungi were isolated (Fig. 9.3). Out of these eleven fungi were members of Deuteromycotina, nine of Ascomycotina, three of Zygomycotina and one form of white sterile mycelium. Individually from *Pinus wallichiana* soils following fungi were isolated: *P. javanicum, T. harzianum, T. viride,*

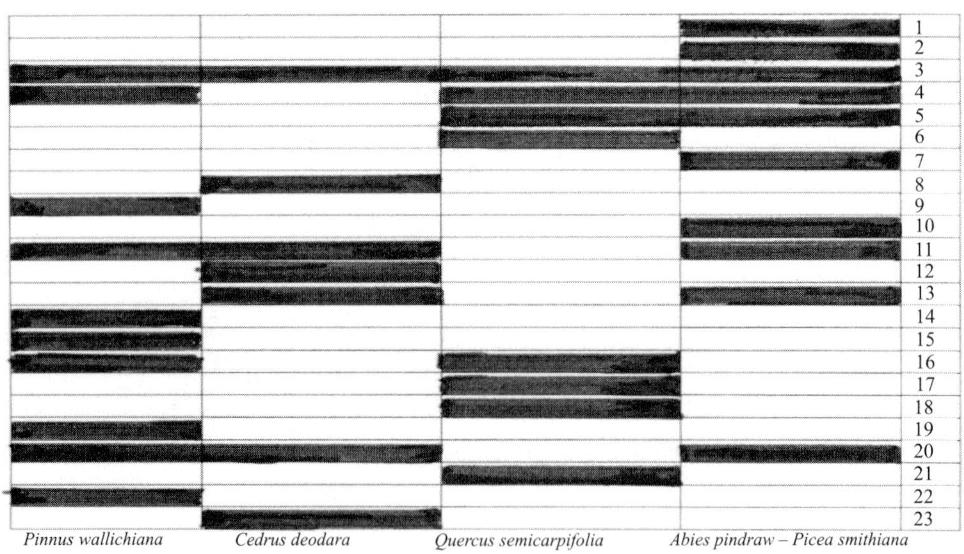

1. White sterile mycelium, 2. *Verticillium* sp., 3. *Trichoderma viride*, 4. *T. harzianum*, 5. *T. glaucum*, 6. *Mucor* sp. 7. *Rhizopus stolonifer*, 8. *R. nigricans*, 9. *Phoma hyvernica*, 10. *Penicillium* sp., 11. *P. jvanicum*, 12. *P. citrinum*, 13. *P. chrysogenum*, 14. *P. biforme*, 15. *Fusarium moniliforme*, 16. *Fusarium* sp., 17. *Cephalosporium* sp. 18. *C. acremonium*, 19. *Aspergillus niger*, 20. *A. luchuensis*, 21. *A. fumigatus*, 22. *A. flavus*, 23. *Alternaria alternata.*

Fig. 9.3 Fungi associated with *M. crassipes* bearing soils in different forest types

P. citrinum, P. biforme, A. flavus, A. nidulans, F. moniliforme, Fusarium sp. and *Phoma hybernica*. From *Cedrus deodara* forest soil, *Alternaria alternata, P. chrysogenum* and *R. nigricans* were isolated. From the *Quercus semicarpifolia* forest soil, *T. glaucum, T. viride, T. harzianum, A. fumigatus, Fusarium* sp., *Cephalosporium acremonium, Cephalosporium* sp., and *Mucor* sp. were isolated. In *Abies pindrow - Picea smithiana* forest soil, *T. viride, A. luchuensis, P. chrysogenum, P. javanicum, T. harzianum, T. glaucum, R. stolonifer, Verticillium* sp. *and Penicillium* sp. and white sterile mycelium.

In *M. crassipes, Trichoderma viride* was isolated from all the four forest types. *P. javanicum*, T. h*arzianum* and *A. luchuensis* were isolated from the soils of three forest types, and *Fusarium* sp., *T. glaucum* and *P. chrysogenum* were isolated from the soils of two forest types. The other species namely *Alternaria alternata, Aspergillus flavus, A. fumigatus, A. niger, Cephalosporium acremonium, Cephalosporium* sp., *Mucor* sp., *P. biforme, P. citirnum, Penicillium* sp., *Phoma hybernica, Rhizopus nigricans, Rhizopus stolonifer, Verticillium* sp. and white sterile mycelium were isolated from the soils of one forest type only.

The soil bearing *M. deliciosa* was collected from forests consisting of *Pinus, roxburghii, Quercus incana, Cedrus deodara* and *Pinus wallichiana.* In all twenty fungi were isolated from the soils of these forests (Fig. 9.4). Out of these seven were members of Deuteromycotina, nine of Ascomycotina and one form was represented by sterile mycelium. Individually from *Pinus roxburghii* forest soil associated with *M. deliciosa,* eight fungi were isolated. These are *Aspergillus luchuensis, A. fumigatus, Cephalosporium aspermum, Mucor microsporus, Penicillium chrysogenum, R. stoloniferum, Trichoderma viride* and a white sterile mycelium. From *Quercus incana* forest soil, seven fungi were isolated. These are *Trichoderma viride, T. harzianum, Fusarium* sp., *Penicillium javanicum, P. biforme, Aspergillus humicola* and *Alternaria alternata.*

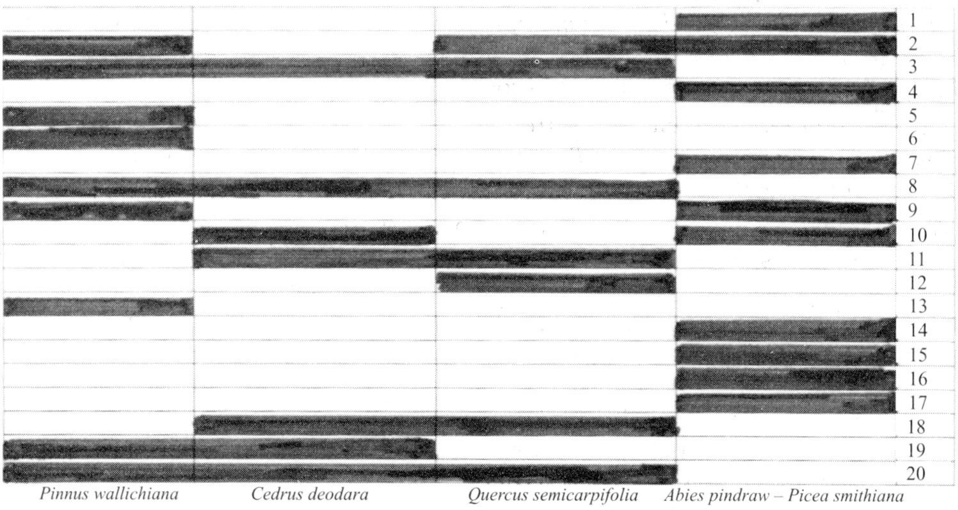

1. White sterile mycelium, 2. *richoderma viride*, 3. . *har ianum*, 4. *Mucor microsporus*, 5. *Rhi opus* sp. 6. *R. nigricans*, 7. *Penicillium stoloniferum*, 8. *P. avanicum*, 9. *P. citrinum*, 10. *P chrysogenum*, 11. *P. biforme*, 12. *usarium* sp. 13. *Cephalosporium* sp. 14. *C. aspermum*, 15. *C. roseum-griseum*, 16. *Aspergillus luchuensis*, 17. *A. fumigatus*, 18. *A. humicola*, 19. *A. flavus*, 20. *Alternaria alternata*.

Fig. 9.4 Fungi associated with *M. deliciosa* bearing soils in different forest types

From *Pinus wallichiana* forest soil seven different fungi were isolated, namely *Alternaria alternata, Cephalosporium* sp., *Aspergillus flavus, Rhizopus* sp., *Cephalosporium rosem - griseum, T. harzianum,* and *T. viride.*

It is clear that out of a total of 19 fungi associated with *M. deliciosa, Alternaria alternata, P. javanicum* and *Trichoderma harzianum* were present only in three forest types, whereas *Aspergillus humicola, Penicillium biforme, Cephalosporium roseum-griseum, P. citrinum, P. chrysogenum* and *Aspergillus flavus* were present in the soils of two forest types. *A. fumigatus, A. luchuensis, Fusarium* sp. *Mucor microsporus, P. stoloniferum, Rhizopus nigricans, Rhizopus* sp., *T. viride* and white sterile mycelium were isolated from the soil of only one forest type.

The soil bearing *M. esculenta* was collected from forests consisting of *Pinus wallichiana, Cedrus deodara, Quercus semicarpifolia* and *Abies pindrow-Picea smithiana.* Twenty-three species were isolated (Fig. 9.1). Eleven were of Deuteromycotina, nine of Ascomycotina, three were of Zygomycotina and one was white sterile mycelium. Fungi isolated from *Pinus wallichiana* forest soil were *P. javanicum, T. viride, T. harzianum, T. roseum, R. nigricans, Cephalosporium roseum-griseum, Fusarium oxysporum, A. fumigatus, A. luchuensis, Cladosporium* sp., *P. funiculosum, P. chrysogenum.* The *Cedrus deodara* forest soil yielded ten fungi: namely *A. alternata, P. javanicum, T. viride, T. glaucum, A. luchuensis, P. citrinum, T. roseum, Fusarium* sp., *Penicillium* sp. and white sterile mycelium. *T. viride, T. harzianum, P. citrinum, F. oxysporium, P. javanicum. A. flavus, M. globosus, P. helicum* and white sterile mycelium were isolated from soils forest of *Abies pindrow-Picea smithiana.*

In *M. esculenta* out of the 22 fungi isolated, *Trichoderma viride* was present in the soils of all four forest types. *P. javanicum, A. luchuensis,* and *T. harzianum,* were present in the soils of three forest types. *Aspergillus flavus, Fusarium oxysporum, Fusarium* sp., *P. citrinum. Trichoderma roseum* and white sterile mycelium were present in the soils of two forest types. *Alternaria alternata, Aspergillus fumigatus, Cephalosporium roseum-griseum, Cladosporium cladosporioides, Cladosporium* sp. *Mucor globosus, Penicillium roseum, P. biforme, P. helicum* and *Verticillium* sp., were present only in the soil of one forest type.

M. semilibera was collected only from *Abies pindrow-Picea smithiana* forest. Total four species viz. *T. viride, A. luchuensis, P. chrysogenum, Alternaria* sp. and an unidentified white sterile mycelium were isolated from the soil of this forest type.

A comparative account of the distribution of different fungal species in the four forest types for *Morchella* species is given in Table 9.2. Indices of similarity calculated for the four forest types of *M. esculenta* indicate that first and second forest types were the most similar (33.33), followed by the third/fourth (24.00); the second/third (13.33), the third/first (10.50) and the fourth/first (8.33) were least similar zones. Indices of similarity calculated for forest types of *M. crassipes* indicate that the first/second forest types were the most similar (35.30) followed by second/third (14.14), the third/first (10.52) and the fourth/first (9.52) were the least similar. In *M. conica,* first/second forest types were most similar (38.09) followed by the third/fourth (22.22) the second/third (11.11) and the fourth/first (10.00) third/first (10.00) were the least similar. Indices of similarity calculated for the four forest types of *Morchella angusticeps* indicate that the first/second forests were the most similar (50.00), the second/third (42.15), the third/fourth and the third/first (31.55) and the fourth/first (12.5) were the least similar forest types. Indices of similarity for the paired *Morchella* species are presented in Table 9.3.

Table 9.2 Indices of similarity calculated for the four forest types of different *Morchella species*

Sr. No.	Paired zone	No. of shared species	Index of similarity
(1) *M. esculenta*			
	First/Second	4	33.33
	Second/Third	1	13.33
	Third/Fourth	3	24.00
	Fourth/First	1	8.33
	Third/First	1	10.50
(2) *M. crassipes*			
	First/Second	3	35.30
	Second/Third	1	14.14
	Third/Fourth	3	33.33
	Fourth/First	1	9.52
	Third/First	1	10.52
(3) *M. conica*			
	First/Second	4	38.09
	Second/Third	1	11.11
	Third/Fourth	2	22.22
	Fourth/First	1	10.00
	Third/First	1	10.00
(4) *M angusticeps*			
	First/Second	4	50.00
	Second/Third	4	42.15
	Third/Fourth	3	31.55
	Fourth/First	1	12.50
	Third/First	3	31.55
(5) *M. deliciosa*			
	First/Second	4	47.00
	Second/Third	5	71.42
	Third/Fourth	1	11.76
	Fourth/First	0	0.00
	Third/First	4	40.00

It is clear from these observations that *T. viride* is associated with all *Morchella* species and all forest types. There is a need to study the interaction on *T. viride* with species of *Morchella,* to see whether it has some stimulatory or inhibitory role to play or not in its growth cycle.

The index of similarity in *Morchella* species for the first and second forest types was highest. It was least for the first and fourth forest types in all the species. The indices of similarity for the first and second forest types were 33.33, 35.30.38.09, 50.00 and 47.00 in

Table 9.3 Indices of similarity calculated for the six species of *Morchella*

Paired Species	No. of shared species	Index of Similarity
M. angusticeps/M. conica	13	53.06
M.conica/M. crassipes	15	58.82
M. crassipes/M. deliciosa	14	65.12
M. deliciosa/M. esculenta	14	63.64
M. esculenta/M. semilibera	4	27.59
M. semilibera/M. angusticeps	4	30.77
M. esculenta/M. conica	14	62.22
M. deliciosa/M. angusticeps	15	43.17
M. crassipes/M. angusticeps	13	59.09
M. semilibera/M. conica	3	18.18
M. semilibera/M. crassipes	4	28.57
M. semilibera/M. deliciosa	4	32.00
M. esculenta/M. angusticeps	15	57.69
M. esculenta/M. crassipes	16	68.09
M. deliciosa/M. conica	13	54.17

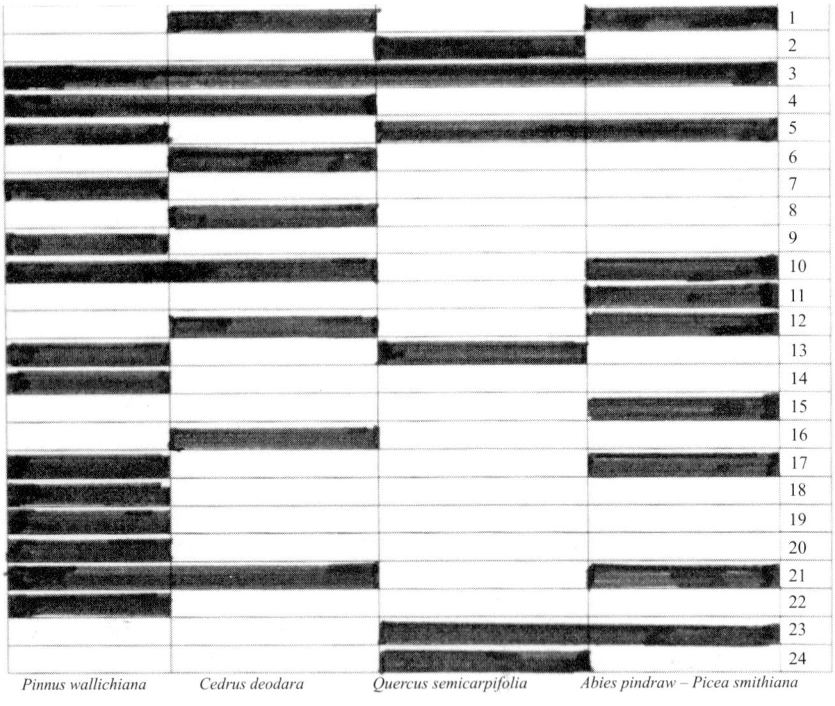

1. White sterile mycelium, 2. *Verticillium* sp., 3. *Trichoderma viride*, 4. *T. roseum*, 5. *T. harzianum*, 6. *T. glaucum*, 7. *Rhizopus nigricans*, 8. *Penicillium* sp., 9. *P. funiculosum*, 10. *P. javanicum*, 11. *P. helicum*, 12. *P. citrinum*, 13. *P chrysogenum*, 14. *P. biforme*, 15. *Mucor globosus*, 16. *Fusarium* sp. 17. *Fusarium oxysporium*, 18. *Cladosporium* sp., 19. *C. cladosporiodes*, 20. *Cephalosporium roseum-griseum*, 21. *Aspergillus luchuensis*, 22. *A. fumigatus*, 23. *A. flavus*, 24. *Alternaria alternata*.

Fig. 9.5 Fungi associated with *M. esculenta* bearing soils in different forest types

Morchella esculenta, M. crassipes, M. conica, M. angusticeps and *M. deliciosa*. The indices of similarity for the first and fourth forest types were 8.33, 9.52, 10.00, 12.5, and 0.00 in *M. esculenta, M. crassipes, M. conica, M. angusticeps* and *M. deliciosa*. The index of similarity was highest (68.00) for *M. esculenta* and *M. crassipes,* followed by 65.12, 63.64 and 62.22 for *M. crassipes* and *M. deliciosa, M. deliciosa* and *M. esculenta* and *M. esculenta* and *M. conica*. These values indicate that *M. esculenta, M. crassipes, M. deliciosa* and *M. conica* shared maximum number of common species and were therefore the most similar in their rhizomorphosphere inhabitants. These values showed that these species of *Morchella* are similar in many characters. The index of similarity was lowest for *M. semilibera* and *M. conica* followed by *M. esculenta* and *M. semilibera* and *M. semilibera* and *M. crassipes.* These low values presumably show that *M. semilibera* is distinctly different from the other species of *Morchella*. (Fig. 9.5)

CHAPTER 10

Ethnomycological Aspects

CHAPTER OUTLINE

- Introduction and Review of Literature
- Observations and Discussion

INTRODUCTION AND REVIEW OF LITERATURE

Ethnobotany is a multidisciplinary branch which attempts to understand plant-human relationships. In fact, if we look back, the study of botany started with ethnic mass who lived in the wild, was nomadic and by hit and trial developed the use of plants. But as it developed into a science, the ancient and traditional knowledge, the folk lores were overshadowed and covered with modern knowledge. But now it has been realized that the traditional knowledge is actually full of wisdom, and has been accumulated and transformed from information to knowledge and wisdom through centuries of experience and cultural refinement. The need of the present times is to authenticate it and make it more utilizable. Ethnomycology is a subfield of ethnobotany, specially dedicated to fungi.

The fungi are a forest resource that is overlooked in ethnobotanical studies (Prance, 1984). The Himalayan hill ranges are inhabited by different tribal populations, often with their distinct way of life, beliefs, traditions, dialects and cultural heritage. The Himalayas have bestowed them with vast, varied and often endemic plant life (Gupta, 1981).

Schultes (1981) regarded the Indian subcontinent to be doubtlessly rich in ethnobotanical wealth, pointing that it is here that many living groups of people, still more or less isolated from the influences of the modern world, continue to live in close association, and vital dependence on their ambient vegetation provide cientists with unparalleled opportunities for profound research. Further, this vast region has other characteristics which make ethnobotanical research highly profitable from both the academic and practical point of view. It supports a most varied and rich flora from the driest of deserts to tropical rain and an extraordinary diversity of altitudes from sea level to the highest mountains in the world. And perhaps of even deeper significance, this region possesses, in the Vedas and a variety of other documents, an ethnobotanical continuum that enables the contemporary investigator to delve into distant past and often to link modern folklore directly with that of ancient people and cultures.

Wild edible fungi have been collected and consumed by the people for thousand of years. The archaeological record reveals edible species associated with people living 13,000 years ago in Chile (Rojas and Mansur, 1995), but it is in China where the eating of wild edible fungi was first reliably noted, several hundred years before the birth of Christ (Aaronson, 2000). Edible fungi were collected from forests in ancient Greek and Roman times and highly valued, though more by high-ranking people than by peasants (Buller, 1914). Caesar's mushroom (*Amanita caesarea*) is a reminder of an ancient tradition still existing in many parts of Italy, embracing a diversity of edible species dominated today by truffles (*Tuber* spp.) and porcini (*Boletus edulis*).

China features prominently in the early and later historical record of wild edible fungi. The Chinese, for centuries, valued many species, not only for nutrition and taste but also for their healing properties. These values and traditions are as strong today as they were centuries ago and are confirmed by the huge range of wild edible fungi collected from forests and fields and marketed widely (Wang, 1987).

It is well known that countries such as Mexico and Turkey, and major areas of central and southern Africa also have a long and notable tradition of wild edible fungi. The list of countries where wild fungi are reported to be consumed and provide income to rural people is impressive.

Wild edible fungi also have medicinal properties, some of which are found in edible species. Wild useful fungi, therefore, contribute towards diet, income and human health. Many species also play a vital ecological role through the symbiotic mycorrhizal relationships that they form with trees. Truffles and other many valuable wild edible fungi belong to this category.

Since the beginning of civilization, people have used plants as food, fodder and medicine. The term '**Ethnobotany**' was first coined by Harshberger (1895) indicating the plants used by Aborigines. Jain (1981) considered ethnobotany as multidisciplinary science comprising many interesting and useful aspects of plant science, history, anthropology and literature. Heiser (1995) considered that ethnobotany is comprised of two fields—first economic botany and then ethnobotany, i.e. first the use and then income becomes the important consideration. He considered ethnobotany as the study of plants in relation to people including both wild and domesticated ones. Therefore, ethnobotany is the 'Botany in service of mankind'. Jain (1995), who is considered as 'Father of Indian Ethnobotany' classified the interaction between plants and mankind into two categories: first is abstract, including taboos, avoidance, sacred plants, worship and folklore, while other is concrete dealing mainly with the material used and the acts of domestication, conservation, improvement or destruction of plants. According to Martin (1995), ethnobotany generally encompasses entire studies concerning plants and describes the interaction of local people with the natural environment. He (2005) defined ethnobotany as "The study of the classification, use and management of plants by the people which draws on a range of disciplines, including natural and social sciences, to show how conservation of plants and local knowledge about them can be achieved". Therefore, it is critical to the growing importance of developing new crops and products, such as drugs from traditional plants. Schultes (1962) defined ethnobotany as "The study of human evaluation and manipulation of plant material substances, and phenomenon including relevant concepts, in primitive or unlettered societies".

Wealth of information on ethnobotany exists in India since the ancient times. Although plants have always been important to people, and the study of the knowledge and use of plants must go back to the beginning of human existence, only in our century this field 'Ethnobotany' evolved as a distinct branch of natural sciences. Actually, ethnobotany has undergone a true revolution in the 20th century (Singh and Aneja, 1999).

The importance of ethnobotanical studies is manifold, as these provide valuable data on: (i) life support services; (ii) new uses not recorded in literature; (iii) new local vernaculature; (iv) distributional areas of new plant drugs; (v) conservation of natural resources; (vi) germplasm stocks of our cultures. All these aspects help in understanding plant-human relationships.

Ethnobotany is multidisciplinary. A subfield of ethnobotany is 'Ethnomycology' which is specially dedicated to mushrooms and has expanding qualities, deep-rooted in religious traditions, writings and indigenous knowledge of mankind. The ethnobotanical studies of the mushrooms have been termed as '**Ethnomycology**' by Wasson (1980). He defined it as, "The study of the use of fungi by the native or tribal people".

Significant contributions to the science of ethnomycology have been made by Wasson and Wasson (1957) and Heim *et al.* (1967). Different ethnic groups utilized fungi in their own ways. Even today primitive tribes and ethnic people utilize plants as well as fungi in their own ways (Charaya and Mehrotra, 1999). Some of the fungi have been utilized as food and for preparing fermented drinks, or for medicinal uses or have been employed in rituals.

In the world, humans have interacted with the fungi throughout history. Ethnomycology is basically the study of the people and fungi and is the recent area of academic interest. These studies can be of great help in suggesting alternative food sources, especially those, which can be obtained by utilizing the raw materials that go waste otherwise. Prance (1984) considered fungi as the forest resources that are often overlooked in ethnobotanical studies. He also studied the native people of Amazoina and revealed a regular consumption and management of wild edible fungi by them.

FAO (2004) lists ethnomycological importance of wild edible fungi: as a source of diet benefits; as a source of income; as a source of delicacy and for health benefits. Ethnomycology includes the cultural, ceremonial and medicinal uses of fungi by people. It was originally dominated by the study of hallucinogenic mushrooms and their cultural significance and very little attention has been paid to the uses of wild edible fungi by people, but as the time passed, the scarcity of food from natural resources and increasing interest of public in new and different food and their medicinal values has build up and expanded the market of wild edible mushrooms.

Ethnic knowledge in mycology is still prevalent in the tribal and local inhabitants of the non-tribal areas (Singh, 1999). If such work is undertaken in the right perspective, the Indian ethnomycological studies can disclose many secrets related to use of fungi in food and medicine by the tribals/local inhabitants to the modern society. Indigenous people are the repository of accumulated experience and knowledge. Unfortunately, this traditional wisdom and the indigenous cultures are threatened by the assault of modern civilization and changes in sustenance economy. It is, therefore, extremely desirable to document the traditional knowledge before it is lost forever.

Even in the present times, in the 21st century, the use of many edible mushrooms in food and medicine in North-West Himalayas is prevalent. From amongst wild edible mushrooms, the most important mushroom from different aspects such as ethnomycological, economic, and social value is *Morchella*, which is still endowed with 'natural habitat' and is also called 'gold mine fungus' because of the price it fetches in the world market.

For ethnomycological information following parameters have been considered:

(i) Habit, habitat and occurrence of *Morchella* species in different habitats.
(ii) The methodology followed by local inhabitants for collection of *Morchella* species.
(iii) Native nomenclature of morels.
(iv) Methodology followed by local people for drying of morels.
(v) Snowfall and *Morchella* production.
(vi) Consumption of morels.
(vii) Traditional medicinal uses of morels.
(viii) Myths associated with morels.
(ix) Marketing of morels.
(x) Conservation strategies.

The standard methodology for recording ethnic data on *Morchella* is given in Annexure - VIII. Observations on all parameters are being presented one by one in the following pages.

OBSERVATIONS AND DISCUSSION

(i) Habit and Habitat

Major crops of morels appear in the Himalayan region during the spring season, i.e. from March–May, after the snow melts. A small crop comes up, however, during the rainy season. Since they appear during two different and distinct seasons, they are called "Spring" and "Rainy season morels".

(a) Spring morels

Morels normally appear during spring season, so they are also termed as 'spring mushrooms' (Plate–Ia). These spring morels are also called "winter morels". They generally start appearing in February and continue till May or mid-June depending upon the area and the incident snowfall. The peak season for their appearance in the Himalayan region is March–May (Shad, 1989; Sharma, 1993; Rathore, 2001 and Rana, 2005). Even in countries like France, morel collecting season has been reported to be from January to June (Delmas and Bunel, 1974).

Since morels appear only after snow starts melting followed by occasional rains, it lends support to the speculations of Schisler and Baker (c.f. Baker and Cook, 1974) that fruiting bodies of morels in spring may be due to their ability to grow at low temperatures to exclude competition.

A distinct trend has been observed in the appearance of morels in the N-W Himalaya. *Morchella angusticeps* (black morels) has been seen to appear first, followed by *M. deliciosa, M. esculenta, M. conica,* whereas *M. crassipes, M. tibilica* and *Morchella simlensis* sp. nov. (white ridged morel) appear towards the end of the spring season. The collections usually start from lower altitudes and gradually proceed towards higher elevations as the season progresses. The colour of species varies from white, yellowish, wheatish to blackish. This is because the snow melts at lower elevations earlier and proceeds towards the higher elevations.

In the tribal area of Himachal Pradesh, called Kinnaur, which has a beautiful and vast fertile valley, the Baspa valley, and where traditional knowledge is still preserved, the fruiting bodies generally appear at high altitude areas. These areas are named indicating melting of snow and the dwellings. These areas are locally called 'Chakka' and 'Goli', the former means 'the place where snow is melting' and the latter 'the place which is far from human dwellings'. The nomenclature changes with the change in valley. In Satluj valley these places are called 'Neols, 'the place near to the dwellings of God', while in Lahaul and Spiti another tribal belt of Himachal Pradesh Himalaya, the forest producing morels are called 'badang-eve, i.e., the places specially created by God, and in tribal valley of Chamba these places are called 'Giri' meaning 'higher altitude areas'.

The fruiting bodies also occur not only in forested areas, but also in the orchards, and fields, and cool and moist places. In apple orchards the creamish fruiting bodies of *Morchella* usually *M. deliciosa, M. esculenta* and *M. conica* appear in abundance after rains and

PLATE–I (a–e) a. Fruiting bodies of the spring morels in the natural habitat; b. Fruiting bodies of monsoon morels. Different habitats of morels; c. Forest areas; d. Burn-site habitat; e. Landscaping habitat (Near surrounding of human dwellings)

thunderclaps. Sometimes the areas around apple trees can have dozens of them. Accompanying morels are usually the fruiting bodies of *Rhizopogon*, which occur in abundance and are also regarded as a delicacy.

Kuo (2002 c) studied the pattern of appearance and distribution of morels in California and observed that black morels are the first to appear followed by yellow and half-free morels.

(b) Monsoon morels

Some species of morels have also been collected during rainy or monsoon months and these are being called here as "monsoon morels". However, during the rainy season only a few species of morels, e.g. *M. esculenta* and *M. deliciosa* and in much less numbers compared to spring morels, appear. Monsoon morels usually come up at lower elevations towards the end of rainy season and are considered inferior to the spring morels by local inhabitants (our analytical studies reveal that they have same nutritive components). The fruiting bodies of monsoon morels are relatively larger in size heavier with water, than winter morels probably because of high moisture content during rainy season (Plate - Ib).

Wakode (1983) collected some species of morels from Maharashtra during the rainy season. Kamal (2005) collected some morel specimens from the Terai region of Uttarakhand. Kaviyasaran *et* al. (2006) collected some specimens of *M. esculenta* from Western Ghats. A good crop of morels was obtained in the months of October and November at Bahu during 1985, at an altitude of 500 m. Enquiries during our forays did reveal their collections at lower elevations (500 - 1000 m) during the rainy season viz. fruiting bodies of *M. esculenta* have also been collected during rainy season of 1972-73 from near Nahan (Altitude approximately 400m) and in September 1982 from Ramshar, near Nalagarh (Altitude approximately 500 m) again towards the end of the rainy season, i.e. September–October.

It appears that in areas not experiencing snowfall, *Morchella* species do come up during rainy season in forests or pastures towards the end of rainy season, i.e. September–October, probably when autumn and winter set in and night temperature falls below normal, which is perhaps congenial for their growth. Personal enquiries at a number of other places in Shiwalik ranges (300 m - 800 m) of Himachal Pradesh confirmed their occasional collection by people during the rainy season, especially towards October and November. When they were shown fruiting bodies of *Morchella* to know if they had ever seen such fruitings, it was revealed at village Haretta in Hamirpur district that fruiting bodies of *Morchella* were more frequent some 50-80 years ago. Some elderly people had collected and consumed them when they were school-going children, during the rainy season. Similarly, collections have also been reported from other places with similar altitude and vegetation.

Gowry (1990) conducted survey on monsoon morels of Himachal Pradesh. She gathered some fruiting bodies of morels in some parts of Shimla and Kullu in rainy season and reported that people consider them to be inferior to spring morels in quality, weight and colour. Afterwards, similar collections of monsoon morels were made by Sharma (2001) while collecting general wild, edible mushrooms of Hamirpur district. She collected fruiting bodies of *Morchella esculenta* from Jolsapar, Hamirpur, and observed that in Hamirpur area morels do come up with other rainy season mushrooms.

Majority of the species of *Morchella,* so far, have been reported from N-W Himalayan region, Jammu and Kashmir (Cooke, 1870; Sydow and Butler, 1911; Ghosh and Pathak, 1962; Batra and Batra, 1963; Waraitch, 1976; Kaul *et al.,* 1976 and Kaul, 1981), and Himachal Pradesh (Sohi *et al.,* 1965; Lakhanpal and Shad, 1986 a, b; Sharma, 1993; Sharma, 2001; Rathore, 2001 and Rana, 2002). J & K, Himachal and western Uttarakhand (Hennings, 1901 and Theissen, 1911) are the main producers of *Morchella* in India (Shad, 1989 and Jandaik and Sharma, 1995).

Morel fruiting bodies occur solitary as well as in groups. The fruiting bodies usually are, black and light brownish or creamish. The black morels were observed to occur solitary or in pairs, while the white ones occurred gregariously, sometimes as many as 20-30 at one place.

Miller (2003) categorized morel habitats into four types: 'Landscaping morels', 'Burn-site morels', 'Orchard morels', and 'Natural morels' (Table 10.3). He defined them as follows: landscaping morels are those which are found in urban settings, i.e. in people's yards and gardens or near by surroundings of human dwellings (Plate – Ie). Burn-site morels (Plate - Id) are those that occur at burnt places (Plate – Ic). Orchards morels occur in orchards and the natural morels in natural or man-made forests. All four types of habitats exist in the Himalayan region, and in all these types of niches, morels have been collected. It has been known for centuries that in the year following a forest fire, morels may appear in great numbers. The morels at burn-sites are generally black morels (*Morchella angusticeps*). Burning of sites is a common practice in some parts of Himachal Pradesh in the hope of selling a good morel crop next season and for raising a good grass cover. This, however, is a very wasteful exercise. This causes great losses to forest wealth, especially in Kinnaur and Chamba districts. The practice of burnt woodlands became a major cause of forest destruction in Germany. To prevent the people from this habit, ultimately it was banned by law (Kaushal, 1986). Even though, it is banned, frequent forest fires engulf every year, advertently or inadvertently.

Kuo (2003) studied the production of burn-site morels from west coast and discussed why there have been frequent forest fires in the west coast. Miller (2003) monitored morel appearance in a Colorado burn area. During 2002, the San Juan National Forest area had number of wildfires, and the following year, i.e. 2003, huge number of morels appeared at the burn-site.

Morels also appear regularly in the 'Orchards'. Apple orchards are a very good morel hunting site, especially the basin around apple trees. Morels occur naturally in diverse habitats like pastures, moist and shady places (Plate–Ia), beneath trees, stones and shrubs and leaf litter, etc. Out of these four habitats, burnt-woodland have been considered to be best habitat of morels.

(ii) Collection of Morels

Previously, just three decade ago, in the Himalayan region, morels were collected by people just as a hobby for home consumption. Now it has become a money earning propostion. All categories of people irrespective of caste, creed and age collect them. It is an important item of commerce in national and international trade. People in villages, shepherds, herdsmen, local inhabitants, young and old women, children make early morning trips to forests and other localities, where morels are known to appear, for collecting them. Even school-going children make quick trips in the mornings to collect them before leaving for school. After school, or sometimes even during school time, children can be spotted in various forest areas collecting morels. The collectors include 40% women, 35% men, and 25% children.

One can easily see a person coming from a hill-top with a bundle of fuel wood on his head and morels in lap or baskets. The fresh collections are carefully kept in small bamboo baskets called 'phil', 'Changer' or 'Chanto'. In Kinnaur, ladies collect them in their 'thumbo' (a fold of head cover), which is the part of their local dress 'dodhu'/'pattu' (shawl) (Plate - II c, d, e, f. h).

In Chamba, the baskets used for collections are called 'Ckakoo' or 'Chaboo' and the part of dress used by ladies for collection is called 'paloo' or 'bhori' formed by shawl (pattu).

In Lahual and Spiti, these baskets were called 'Kirti'; 'Chaboo'; 'Perra'; 'Panja' and the part of ladies' dress for keeping collections is called 'Katar' or 'Dhupo'.

In the tribal areas, two categories of people who are migratory and traditionally have been rearing cattles and mainly live in the forests called 'gadhi' and 'charani's' visit forest very often. They have their house at low and high altitude areas (Plate - II g). Their dwelling place is called 'dogri', so while on the way to 'dogri' they collect the fruiting bodies.

Morel collection is a hectic job and requires a lot of physical exertion, effort and passion. The local inhabitants take keen interest in their collection because of the income it provides. They follow different methods and ways of collection. To a novice, company of an experienced morel hunter is always fruitful. The area needs to be screened very carefully with aptness. Once spotted, in a niche the nearby area will also normally have morel fruiting bodies. Morels usually tend to blend with the surrounding ground cover. Careful treading of the ground also helps in avoiding stepping on the snakes, which may be near the fruiting bodies, occasionally.

Around 25-30 years back, *Morchella* fruiting bodies were collected along with *Viola* (Plate - IIIc), a medicinal herb, commercially called 'banafsha', an important ingredient in Indian System of Medicine, but people now have been focussing their attention on *Morchella* alone. *Viola* collection has declined due to low monetary returns. Still people in Kinnaur, while collecting morels, do collect other medicinal herbs, as well as those which are locally consumed by people as vegetable or/and used as condiments.

In Kinnaur it is a common belief that snakes are present where morels grow. We have also observed snakes near morels at quite a few sites. But the villagers consider these snakes harmless. The people believe that before embarking a morel-collecting trip, if one prays to the local deity, the trip will be fruitful.

Simultaneous collections of morels and medicinal herbs are practiced in Pakistan as well. In Swat district (Pakistan) about 500 families are involved in medicinal plant and morel collection and they collect about 5000 tonnes of medicinal plants and morels annually (Chaudhary *et al.*, 2000). Similarly, in Utror - Gabral valleys of Pakistan, it has been observed that villagers collect morels along with medicinal plants. They hand over their collection bag (including morels + medicinal plants) to local dealers of morels and medicinal plants (Hamayun *et al.*, 2003). In Himalayan region of India also the collections are usually sold to agents engaged by local dealers or by dealers from Delhi and other cities.

PLATE–II (a–h) a– Drying of morels by making garlands. b- Garland of morels with dry fruits made for festive and auspicious occasions. c, d f– Tribal people showing their morel collections. g– Local women bringing her morels collection while visit to the forest. e, h– Collector exhibiting dried morel garland.

(iii) Native Nomenclature

In the western world the morels are known by different names such as 'pine cones', 'sponge mushrooms', 'dry land fish' as 'merkel' and commonly as 'brains' because the grooves impart them brain-like appearance. In the Indian market, the morels go by a common name, the 'guchhi' and same is the common trade name. In the Himalayan region, the morels are known by different names in different valleys.

Information on native names has been gathered from almost all places in Himachal Pradesh. *Morchella* is known by different names following differences in dialects. In majority of the areas in and around Shimla, the common name of *Morchella* is "Chiaun" which becomes "Kaiun", in Kullu valley. In some areas of Kullu and Mandi, they are also called "Dunghlu", and "Chunchuru".

Different species of *Morchella* are white, yellowish, wheatish, or blackish in colour. Their colour is compared with different cereals and millets and they are called 'Chawal (Rice) ke dunglu', 'challi (Maize) ke dunglu', 'Gehun (Wheat) ke dunglu' and 'Kode (*Elusine*) ke dunglu'. These cerealized names have been given to the morels because people believe that the fruiting bodies of *Morchella*, depending upon their colouration, are made up of the respective cereals and the taste varies according to the colour and is like the cereal, it is named after. They consider white and wheatish fruiting bodies to be most delicious.

In the tribal belt of Kinnaur district the local name changes, as there is change in the valley. In Baspa valley, it is called 'Rangmooch', 'Rangmu', 'Rangmutz' and Rangmuto'. 'Rangmooch' – 'Rang' locally called 'kanda', means high altitude place and 'mooch' means 'anything growing without seeds'. While in Satluj valley, it is called 'Jangmooch', 'Jangmo', 'Jangmutz'. In Kalpa area white guchhi is called 'Ban-Jangmoo' and black one as 'Rok', whereas at Rarang and Sipilo it is called 'Kanangujri'. Small kids find it difficult to pronounce, so they call it 'anurag'.

In Chamba district it is locally called 'Chamkeru', 'Chamkenu', 'Chnau', 'Chamkaid', and 'Kankan-egdu', etc., while in Lahaul and Spiti it is called 'Dhuna' and 'Dhunglu', etc. Local and vernacular names are given in Table 10.1.

(iv) Drying of Morels

The fruiting bodies have around 80-85% moisture. Hence, they need to be properly dried. Traditionally, drying has been the easiest, fastest, and long lasting method of preservation. The fresh fruiting bodies are first trimmed to remove the dirt, debris and bad spots. If fruiting body is a little bit slimy or is weak and watery, then it is horizontally sliced and split after washing.

Fruiting bodies are dried in the sun and stored for future use. Like the drying of maize cobs, these are spread on 'slate' roofs during clear days. They are also threaded in a string making 'morel – garlands' (Plate – IIa). The garlands are then dried in the sun or by hanging them above the fireplace where fuel wood is burnt for cooking. At last the specimens to be preserved for marketing are wrapped in paper bags and those kept for home consumption (food or medicine) are hanged in the kitchen. Kuo (2002 b) also suggested the same methods for the preservation and drying of morels.

Table 10.1 Local vernacular names of *Morchella* species

Sr. No. Area visited	District	Local, Vernacular name
1. Kalpa, Roghi, Chinni, Kashmir, Kothi gaon	Kinnaur	Jangmutz, Jangmoo, Jangmooch Ban-Jangmoo, Rok.
2. Nichar, Nigulsari	Kinnaur	Jangmu, Rangnu
3. Sangla, Rakchham, Dangdangshi, Khrogla, Karcham, Chitkul, Mustrang, Chanshu, Shoung	Kinnaur	Rangmuto, Rangmutz, Rangmooch Rangnu
4. Rarang, Sipilo,	Kinnaur	Kanangujri
5. Ropa, Rally	Kinnaur	Jangmoo
6. Rushkalang, Kamru	Kinnaur	Rangmutz, Rangmooch Rangnu
7. Dubling, Rispa, Yangthang, Hango, Nymgah, Jangi, Thangi, Lippa, Arrang	Kinnaur	Jangmutz, Jangmoo, Jangmooch Ban-Jangmoo
8. Churah valley	Chamba	Chamkedu
9. Himgiri valley	Chamba	Chamkenu
10. Bhandal valley	Chamba	Chanu
11. Bharmour valley	Chamba	Chamkedu
12. Tisa valley	Chamba	Chamkedu
13. Pangi valley	Chamba	Bhuntu
14. Mayal, Toth Tinen, Ghar	Chamba	Duna
15. Pattan valley	Lahaul & Spiti	Dhunglu, Chunchru

Some precautions, which need to be undertaken while drying morels are:

- The fruiting bodies should not be washed before drying because this will make them wet and soft.
- The fruiting bodies should be dried near heat and air.
- If there is any doubt of bugs inside the fruiting body, it should be cut horizontally and then dried in the sun.

(v) Effect of Snowfall on Morchella Production

There appears to be a distinct, definite and direct relationship between *snowfall* and *Morchella* production. Heavier the snowfall in winter, better the crop in spring, is the common observation, provided there is good and continuous precipitation in the subsequent months. Low temperature seems essential for fulfilling chilling requirement of primordia. Mostly the fruit bodies start emerging as the snow begins to melt but for better crop, rainfall is also essential. Thus, it appears that chilling perhaps is essential in the production of *Morchella.* This lends support to the speculation of Schusler and Barber (c.f. Babar and Cook, 1974) that the fruiting of morels in spring may be due to their ability to grow at low temperature due to the exclusion of competition. This stands fully corroborated by our observation of almost 25 years. Whenever there was heavy snowfall in winter months, there was luxuriant crop of morels in the spring and vice versa. To cite some specific examples; in the years 1982-83, when there was unusually heavy snowfall from December to March followed by rains in the spring of 1983, there was excellent crop of *Morchella.* In 1983-84, the snowfall had been heavy, but precipitation little, so that *Morchella* species were comparatively less abundant and also smaller in size. In 1985-86 snowfall and rainfall, both have been scanty and there was further decline in the production. 1987-88 were drought years and the production was very low but the prices almost three times. In 1988-89, there has been very heavy snowfall and *Morchella* started appearing. But the rates were slightly lower than the previous year. 1989-90 was an excellent collection year. The year in between 1998-99, had good and bad morel seasons. In 1999, there has been very scanty snowfall, hence the morel crop has been very less productive. In 2004, there was scanty rainfall and morel crop was also less but in 2005 there was heavy snowfall and a bumper morel crop. In 2006 there was scanty snowfall and less morel production, and in 2007 there was heavy snowfall and heavy morel crop production.

Morchella spp. may occur during rainy season as mentioned earlier, but they do not occur in such abundance as in spring. They are sporadically distributed at lower elevations even where chir forests are well represented. The same chir forests in the vicinity of Shimla, at about 1600 m, produce good crops of morels. However, the main season of their appearance is the spring season and main areas, the places which experience snowfall. Even in such areas, the places which experience heavy and well-distributed snowfall have higher production figures for morels.

Table 10.2 Local vernacular names of *Morchella* species (Shad, 1989)

Sr. No.	*District*	*Local Name*
1.	Kullu	Dunglu, Chunchuru, Kiaun
2.	Shimla	Chiaun
3.	Mandi	Chunchru, Dunglu
4.	Chamba	Chamkaid, Kankanegdu, Chamkenu
5.	Kinnaur	Jangmoo
6.	Sirmour	Chiaun, Guchhi
7.	Solan, Hamirpur	Guchhi

Observations contrary to these have been made by Jacobson (2003). He observed that there is no relationship between the amount of winter snow and the growth of morels. He does find that April and early May rain followed by warm but not hot weather encompasses the morels to grow. If rain stops before mid-April or a hot spell arrives, the morels dwindle. If the rain continues, but the weather gets hot causing the soil temperature to rise, the morels dwindle. So, even if the ground is moist and the soil temperature is too high, morels will stop fruiting. Mostly, early morels arrive right on time, but the really giant morels do not grow unless the moisture continues and the soil temperature stays down. So, there exists no relationship between morels and winter snow. However, our observations in the Himalayan region suggest a definite relationship between snowfall and morel production and our observation span almost 25 years.

Table 10.3 Habitats of *Morchella* species

Sr. No.	*Habitats*	*Percentage of occurrence*
1.	**Burnt woodland in forests**	40-45
2.	**Landscaping**	5-10
3.	**Natural Places** (Leaf-litter; near water resources; beneath trees, stones, shrubs, pastures and grasslands, etc.)	20-25
4.	**Orchards** (apple and apricots, etc.)	15-20

(vi) Consumption of Morels

The commercial aspect of morels has reduced their home consumption and people prefer to sell them and earn money (Table 10.4). The black fruiting bodies have been considered more delicious than white and wheatish ones. These are now served only on some festive occasions, to distinguished guests and marriages that too in some rich families. In some marriages at typical Kinnauries and tribal belt of Chamba, the bride and groom have 'garlands' of Guchhi as 'Jaimala' (Bridal garland). But now due to scarcity of fruiting bodies and increased commercial value of morels, the garlands are made of dry fruits especially 'chilgoza', and dry fruits in between some fruiting bodies of morels are also placed (Plate - IIb). People usually knead the garland with black colour morels because as they are first to appear, they are regarded as a sacred gift of 'Mushroom God'. Recipes for the consumption of morels are given separately.

Table 10.4 Per cent use of morels for different purposes

Sr. No.	*Use*	*Percentage*
1.	As food	0
2.	As medicine	5-6
3.	For sale	90-95

(vii) Use of Morels in Traditional Medicines

The local inhabitants in remote areas depend upon folk medicines and household remedies to a great extent. The prevalent practices of indigenous herbal medicines including fungi have descended from generation to generation and include cure of both simple and complicated diseases. The knowledge is usually passed from one generation to another among the families of the village medicine men. Of all the edible mushrooms of N-W Himalayas, morels have been traditionally used for the cure of ailments such as pneumonia, fever, cough and cold, dehydration, stomach pains, arousing sex appeal and for pregnant and lactating mothers. This is also known to cure all the respiratory ailments. Whole fruiting body of the morels (both stipe + pileus) is used to cure various ailments and generally they prefer to use black coloured fruiting bodies of morels. People have common belief that the medicinal properties depend upon the age of the fruiting body; older the fruiting body, more potent it is.

While doing PRA it was noticed that villagers usually pay no visit to local dispensaries but they make their own formulations to cure patients. All the medicinal uses are given in Table 10.5. There is scanty data available on the traditional medicinal properties of morels. Rana (2002) for the first time gathered original and indigenous data on the medicinal uses of morels from district Kinnaur.

Table 10.5 Traditional use of *Morchella* species in medicine

Sr.No	*Ailment*	*Method*	*Use*
1.	**Stomach pain**	Boiled water of the fruiting bodies is given to the patient or fruiting bodies grinded in water or raw milk.	Relieves from pain.
2.	**Pneumonia**	Decoction is prepared of old fruiting bodies of guchhi as 'Garam kadha' is given to the patients.	Cures pneumonia.
3.	**Dehydration/ blood stools**	Boiled water of fruiting bodies by adding little salt/sugar.	Recovers water loss of the body.
4.	**Respiratory problems**	Decoction is prepared by adding 2/3 fruiting bodies to 3 glasses of water. Boil till water remained ¼.	Cures all the respiratory problems.
5.	**Fever, cough and cold**	Boiled hot soup or raw guchhi is eaten by adding little salt and desi-butter.	Heavy sweating is there, which lowers the body temperature.
6.	**Healing of wounds**	Paste is prepared by grinding the fruiting bodies with desi butter and then applied on the wounds.	Heals the wound fastly as it provides warmness to the wound.
7.	**Pregnant women**	Soup, 'subzi of guchhi' (dry vegetable) and decoction is given to the ladies.	Provides strength, warmness to the body and considered very nutritious.
8.	**Lactating mothers**	Decoction and 'subzi of guchhi' (dry vegetable) is given to the ladies.	It is highly nutritious and provides energy to the body which indirectly is given to the baby.
9.	**1-12 months kids**	Fruiting body is grinded with honey/raw milk – 3 spoons/day; soup once per day.	Growth of kids.
10.	**Asthmatic patients**	Simple frying of fresh or rehydrated fruiting body or decoction is given to the patient.	Controls proper functioning of lungs.
11.	**In acne**	Paste is prepared while grinding in water.	Due to warm nature it will dry that and cures it.
12.	**Sex appeal**	Decoction is prepared in apple wine.	Provides warmness to the body.
13.	**Weakness**	Decoction is prepared by adding milk and while serving add 2 spoons of honey.	Provides strength to the body.

(viii) Myths Associated with the Morels

There are many myths associated with mushrooms. Almost similar myths are associated with morels. *Morchella*, it is believed, is easily spotted only by a few gifted or lucky persons.

It is also believed that morels appear when rainfall follows thunder and lightning. Heavier the thunder larger the crop. It is believed that the soil cracks due to lightning and the fruit bodies emerge out through the cracks. The experimental evidence to support or refute these myths is, however, lacking, but circumstantial evidence confirms the appearance of morels in plenty only after it rains. And this has scientific reason in that the buttons will immediately sprout up after receiving suitable moisture through rain, just as it is true for the appearance of fairy rings of mushrooms and usually rain is accompanied by thunder. People also believe and it is corroborated the world over that morels appear more on burnt up soils. Soils rich in calcium have been reported to bear more morels. Burning probably enriches soil with calcium. In many other countries similar practices have been followed by people to harvest rich crop of morels. This practice of burning woodlands became a major cause of forest destruction in Germany. To prevent the people from this habit, ultimately it had to be banned by law (Kaushal, 1986).

Kaushal (1986) also reported that "Gaddi people" in Palampur area of Himachal Pradesh believe if one carries some morels along with him, one is able to collect more morels on the corollary of "like attracts the like". People in many places believe that one gets more fruiting bodies of morels in the morning hours of the day. This we found to be true but actually depends upon the aspect of the hill explored. If one is exploring a hill facing the east, the morning sun penetrates deep and one is able to see through beneath the bushes where morels usually occur. Further, the morning sun is also not very bright as the midday sun is, so in such situations one walks with back towards the sun and the glare of sunlight does not distract eyes from concentrating on spots for morel collection. We have also noticed that morels are collected while climbing up than down, because going up one paces slow, therefore, spots morels, while coming down one glides fast and is not able to concentrate.

The collecting trips may not be productive always, especially for researchers, who are not very familiar with morel bearing sites as are local inhabitants who keep such habitats marked and therefore, their trips are mostly productive. We have experienced that one may wander the whole day and collect nothing. On April 22, 1982 we were a group of eight people, wandering in search of *Morchella* species, especially *M. semilibera,* of which we had just one fructification collected the previous year from the same area in Chazpur (H-P) at an altitude of 3000 m. We were in the forest for almost eight hours but could not get even single specimen of this species. Disheartened we decided to go back, but then we spotted one fructification and in few minutes we collected around 80 specimens of *M. semilibera.* The sight and delight were beyond expression and the tiredness disappeared with "sumptuous" collection immediately.

Many 'guchhi hunting theories' are prevalent with local inhabitants of N-W Himalaya, which are associated with the appearance and collection of *Morchella.*

(a) Other myths associated with morel appearance and collection

- It is believed that one should leave at least one fruiting body at some collecting sites for 'guchhi Devta' (Morel God), so that he returns two to three times more morels. This will ensure the eruption of two-three times greater number of fruiting bodies next year. In fact, leaving a fruiting body is a scientific method as it ensures inoculum for next season. Crushing the fructifications at a site may prove more useful for inoculum multiplication *in vivo.*

- Old apple orchards and fields are considered to be blessed by God for a longer period, hence, they have more production of guchhi than the newly raised ones.
- For the appearance of very good crop, people worship 'Rain God', so that there will be more thunderclaps and rainfall and they can fetch good amount of fruiting bodies.
- It is believed that once the fruiting body is touched, it does not increase in size, if it does, then the stalk slowly and slowly disintegrates from the base.
- It is believed that morels are a gift of nature specially created by God, so it will be available away from the human dwellings.
- Snakes and morels have some associations, so that where morel fruiting body appears, nearby surely there will be a snake.
- People believe that bigger sized fructifications of morels contain snakes inside them. Therefore, the morel hunters first of all make a movement in the pileus, and then after satisfaction, remove it from the substrate.
- If morel hunters go empty handed or without any special preparation for collection, then there will be more chances of the availability. But if prior preparations of the collections are made, then the entire trip may go waste and one may return without single collection.
- While going for collections, the same morning, if one has to fetch good number of fruiting bodies, then he/she must give fresh milk and incense to the nearby water bodies. This will make the God happy and He will help in collection.
- If at a particular place morel hunters fetch good amount of fruiting bodies, he/she should not disclose the secret to anybody, not even to the family members. During next year he can get large amount of fruiting bodies in the same place.
- People commonly believe that if once spotted and seen, the fruiting bodies do not increase in size. Therefore, it has to be plucked immediately from the substrate.
- People relate the collection of guchhi with diligence and patience, because if you are hard-working then only you can toil and concentrate upon their collection.
- Sometimes people carry one or two morels along with them in their shawls or in cap without disclosing, and they collect good amount of morels.
- People believe that to collect a good lot of morels, one should go covered with a blanket.
- If one has a dream of morels he/she is able to collect them surely.

Hamayun and Khan (2004) also recorded some of the myths associated with the collection of morels in Pakistan. They consider the morel collection as mythical one and people in that area believe morel collector as fortunate person. Sometimes the collector even puts antimony in his/her eyes, as there is a myth that such collectors will collect more morels.

(ix) Marketing of Morels

Morchella, at present, is the most prized and priced mushroom having highest market value. Morels are collected in bulk every year from the wild and transported to different markets after proper drying and grading.

Morchella, after collection from the field, exchanges many hands. Local or the primary collectors do not have licence, so they sell the dried or fresh samples to local grocers or display these for sale in local fairs (Plate – III a-f). Most of these are the agents of businessmen and sell them to 'Middlemen' who supply to trading centres or export even directly (Table 10.6).

The middlemen buy the dried collections @ Rs. 800-1,000/kg while they sell it in the market @ 5,000-6,000/kg and it reaches the consumer class @ 8,000-9,000/kg. Obviously, the consumer has to pay three or four times more than the price realized by the collector. The huge share of the middlemen can be avoided easily by providing properly channelled system through morel cooperatives, nationalization of morel collection and by making some easy licensing policy. An estimated 70 ± 5 tonnes of dry morels are exported every year from India in the international markets. Himachal Pradesh, Jammu and Kashmir and Uttarakhand are the main exporters of morels in India (Jandaik and Sharma, 1995). The main species exported are *Morchella angusticeps, M. deliciosa, M. esculenta,* and *M. conica.*

Table 10.6 Marketing channels and price of morels

Sr. No.	*Marketing channel*	*Rate*
1.	Local village grocers/fairs	@ Rs. 2,500-3,000 per kg
2.	Through middlemen (contractors)	@ Rs. 1,500-2,000 per kg
3.	Directly to main market, i.e. Amritsar, Delhi and Pathankot, etc.	@ Rs. 5,000-6,000 per kg

Approximate business of morels from different valleys of district Chamba is Rs. 80–90,000/ season. Whereas approximate business in different valleys of Kinnaur is Rs. 90-1,00,000/ season. There are villages named 'Janiya' and 'Ramni' in Baspa valley where there is largest production of morels every year. Every house sells around 10-15 kg/season and the annual return only from one village is Rs. 1-1.5 lakhs/season. In Lahual & Spiti, the business is less as the production and the collections are not so good. The main trading markets for the selling of guchhi are Pathankot, Amritsar, Delhi and Bombay. From these centres the morels are exported to France, U.K., Switzerland, Dubai, Germany, etc. and their export range from 150-170 millions rupees (Iqbal, 1993). Thus, the exporters are the main beneficiaries followed by middlemen.

i) ***Storage of morels***

Morels need a lot of care in storage as their quality deteriorates very quickly. The prime successful technique is to keep morels dried and cool with a little ventilation. They are kept in closed chambers, mainly tea crates, as in the open they may absorb moisture and start decomposing. Occasionally, insect larvae may infest the morels during storage. In such cases, morels are properly fumigated with insect killing tablets in separate chambers. Sometimes morels get spoiled by mosses, which grow on morels in hot and humid climate. The spoiled morels are generally washed by the traders with water but it deteriorates the quality of the morels, as they turn black.

PLATE–III (a–f) Marketing (a –f). Photographs of local inhabitants trading in morels and a medicinal herb *Viola* **(V).**

Marketing chain of morels is given below:

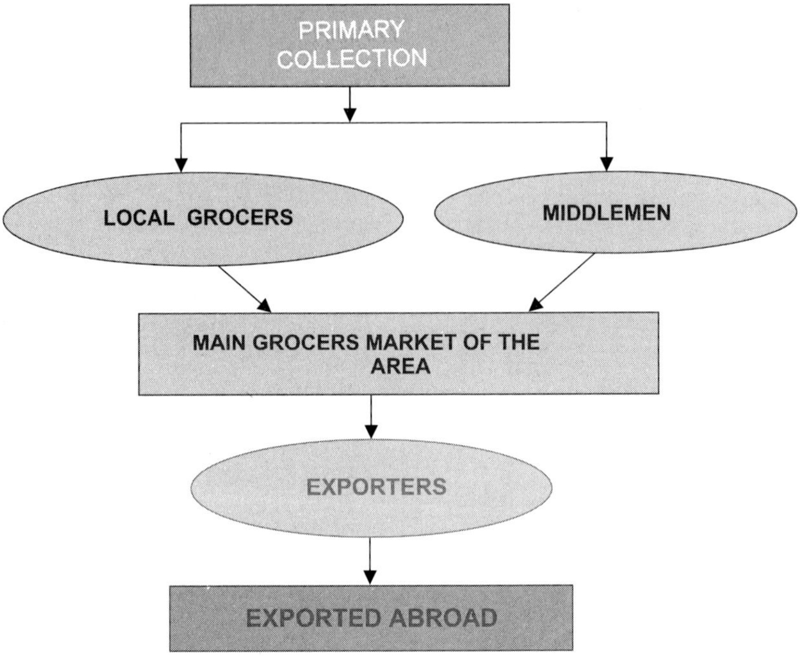

The main causes of morel spoilage are:

- Hot and humid conditions cause a great loss to morels, as hot conditions and increased moisture can make the morels susceptible to insect attack and fungal growth.
- When fresh morels are kept in airtight plastic bags, spoilage takes place as a result of suffocation.
- While growing, the morels get injured due to storms, running water or a piece of wood, etc. Such morels many become very hard and turn black after drying and are considered to be of inferior quality.
- Insects and ants also cause great damage to morels.

ii) Adulteration

The morels are graded on the basis of their colour and size. Large-sized black morels are considered the best ones. They are light in weight as these are hollow from inside. Due to this, to increase its weight, the people sometimes insert iron nails inside. To check adulteration, only the cut pieces of mushroom, which don't have stem basis, are demanded by importing countries. Sometimes villagers even collect *Verpa* and *Helvella* as *Morchella* spp. and mix them with the fruiting bodies of morels. Buyers may or may not have any specifications for morels so they buy entire collection of mushrooms also.

Shad (1989) conducted sociobiological survey on the production and trade of guchhi in Himachal Pradesh (See Chapter - 11). An estimated 50 ± 60 tonnes of dry morels are exported every year from India to the international markets. The socio-economic profile of morels collecting populations reveals that economically weak sections of society and people with small landholdings are primarily engaged in the collection of morels. Rathore (2001) also made similar studies on the production of morels from different forests of Chamba. She also calculated the approximate business of morels per season from different valleys. Pathania and Vashist (2001) discussed the income and employment generation from agribusiness – mushroom farming in H-P. According to them, India is a traditional producer of native wild mushrooms (like Dhingri and guchhi) and is a major exporter of these varieties to Switzerland, France and Germany.

Latif *et al.* (2004) also gave the potential and market status of morels in Pakistan. They gathered the data for the processing of morels, i.e. how the primary collectors sell their collections to local shopkeepers and then from these shopkeepers to middlemen and after that these dried samples are then supplied to main trading centres like Lahore, Karachi, and Rawalpindi, etc. Similarly, Hamayun and Khan (2004) conducted studies from Swat district of Pakistan on the marketing of morels. The main species exported are *Morchella conica* and *M. esculenta.* The collected morels from the forests are finally exported to France, Belgium, Switzerland, Austria and Germany.

(x) Conservation Strategies

There is a feeling among the people that *Morchella* sporophore does not increase in size, once it is spotted. Unfortunately, this belief and high market price has resulted in unscientific and merciless collection of *Morchella*, so that even bits of material are removed from the natural habitat. The villagers collect even small bits of the tissue of morels from the soil, thinking that if the whole fruiting body is plucked, more crop will be there in the next year as this way more space becomes available for fructifications to appear, which is not correct. Ruthless cutting of forests has also resulted in decreased morel production because there is definite relationship between morels and forests, and if this trend continues it will not be surprising that *Morchella* species become extinct in very near future. This is further supported by a report (Buscot and Roux, 1987) and our observations (Lakhanpal *et al.*, 1991) made during the last decade that morels do have some mycorrhizal connections. So, if the plants go, they also disappear simultaneously. Therefore, caution has to be exercised with regard to preservation of forests and it would also be advisable if collectors leave a few fruiting bodies at the site, as some inoculum is ensured for next season. This practice is justified by the experience of the people over the years. We met people during our forays who have been harvesting natural crop of *Morchella* separately at particular places by crushing few fruiting bodies at the site of collection. To mention just one, a person in village Jarol Tikker near Narkanda (Shimla) has been harvesting *Morchella* at one specific site for the last 20 years by this technique. Another person informed us that he collected soil from morel bearing sites at an altitude of 2200 ft and then left it for almost a year in trays, at his home located at 1500 m, regularly watered it and obtained morel fruiting bodies the next season, in plenty. But such soil failed to produce any morels in the subsequent years. This is explainable on the basis that such a soil was having some rhizomorph or mycelial bits from which emerged fruiting bodies in the next year but such inoculum was absent in subsequent years and hence no fruiting bodies. Let the 'no inoculum' stage not come in the forests by judicious planning.

So far no measures have been undertaken for conservation of morels. Since, it is still a resource from the wild, over exploitation can lead to the total depletion and disappearance of this resource. Hence, it is proper to suggest/devise conservation strategies.

However, there is some traditional wisdom used for conservation. The traditional morel hunters pick morels very delicately, cutting the stipe with knife and leaving the base and underground mycelium which serves as inoculum next season. There are other collectors who leave 1-2 fruiting bodies at some specific sites, believing that this mother guchhi will surely produce many daughter guchhi's next year. Next year when these sites were visited, each site has a good crop of *Morchella* waiting with 20 or more fruiting bodies.

Such traditional picking methods ensure propagation and can help in morels not becoming endangered. Certain other conservation strategies, which have been followed elsewhere, can also be promoted in the Himalayan region such as:

- Do not pull the morel as it disturbs the mycelium.
- Gently cut the fruiting body from near the base, so that mycelium remains intact and produces fructification next year.
- Do not pick all the fructifications; leave some fruiting bodies at each site, so that the inoculum for next crop is ensured.

Shad and Lakhanpal (1986 a) discussed similar propagation methods for the further propagation of fruiting bodies of morels. They also suggest that this wonderful natural product, which yet awaits artificial cultivation on commercial scale by scientists, may be completely exhausted before the cultivation process is standardized. They also mentioned that ruthless cutting of forests has also resulted in an imbalance between morels and forests. Similarly, Kuo (2002 a) also gathered data on some misconceptions on picking morels, by which we can elude some conservation methods for the propagation of morels.

CHAPTER

11 Sociobiological Aspects

CHAPTER OUTLINE

INTRODUCTION

Himachal Pradesh possesses a diverse range of climatic zones from subtropical to temperate and alpine. Each zone is, therefore, endowed with a specific vegetation type. The diversity in vegetation results in a still greater diversity in fleshy fungi. Species of *Morchella* also exhibit diversity with change in vegetation and altitude. They commonly occur in abundance in the coniferous forests. Majority of the people in Himachal Pradesh live in rural areas and practice agriculture, horticulture and animal husbandry. In many areas people undertake morel collection as an income generating activity.

India is the major exporter of 'Black Himalayan Morels' along with Pakistan and Italy. An estimated 50-60 tonnes of dry morels are exported every year. The lucrative business is in few hands who export the fungus to Europe where it sells at an average price of around US$ 150–200 per kg. (Kaushal, 1986).

Himachal Pradesh is presently exporting large quantities of morels in and outside India. In Himachal Pradesh morels occur in areas where means of communication have been provided in recent past. Most of these areas are remote. The real potential of morels as a money earner has been realized only during the past 25-30 years and, therefore, these have come to play a major role in the socio-economic status of the people of these areas. In this chapter, data obtained from studies conducted to understand and investigate social and economic status of morel collecting people, and to record production and trade practices of morels in Himachal Pradesh is discussed.

THE PEOPLE

Majority of the people in Himachal Pradesh live in rural areas. They practise and earn their livelihood directly or indirectly from agriculture and animal husbandry. The villages are generally self-sustaining units and the people are firmly attached to the joint family system. The steady and self-reliant woman of the Pradesh enjoy a greater degree of freedom than their counterparts elsewhere in India and they play a major role not only in the indoor activities but also in the outdoor activities as well.

The population and its density per sq km. by district is presented in Table 11.1. It is clear from the table that the lowest density is in the cold deserts of Lahaul and Spiti and Kinnaur, where the vegetational cover is also almost negligible. Rural and urban population statistics as per the 2001 census are also given in Table 11.1. The figures indicate that the majority of the people live in rural areas of the Pradesh.

AGRICULTURE

About 76 per cent of the population depends directly on agriculture. Taking also into account those people, who indirectly depend upon agriculture, 92 per cent of the rural population drives its livelihood from this primary sector. The income from the agriculture sector is about 50 per cent of the Net State Domestic Product. Most of the areas in Himachal Pradesh are not available for cultivation due to the difficult hilly terrain, which is only available for raising forests, pastures and grazing lands. Only about 11 per cent of the total geographical area is

Table 11.1 District-wise area, total population - rural and urban and density

Sr. No.	District	Area (sq km)	Total Population	Rural Population	Urban Population	Density
1.	Bilaspur	1,167	3,40,885	3,18,934	21,951	292
2.	Chamba	6,528	4,60,887	4,26,345	34,542	71
3.	Hamirpur	1,118	4,12,700	3,82,494	30,206	369
4.	Kangra	5,739	13,39,030	12,66, 745	72,285	233
5.	Kinnaur	6,401	78,334	78,334	-	12
6.	Kullu	5,503	3,81,571	3,51,478	30,093	69
7.	Lahual & Spiti	13,835	33,224	33,224	-	2
8.	Mandi	3,950	9,01,344	8,40,362	60,982	228
9.	Shimla	5,131	7,22,502	5,55,269	1,67,233	141
10.	Sirmaur	2,85	4,58,593	4,10,923	47,670	162
11.	Solan	1,936	5,00,557	4,09,362	91,195	259
12.	Una	1,540	4,48,273	4,08,849	39,424	291
	Himachal Pradesh	55,673	60,77,990	54,82,319	5,95,581	109

Source: Census of India 2001
Series - 7, Himachal Pradesh, Primary Census Abstract.

available for cultivation. In order to get optimum yield per unit area, per unit time, the Department of Agriculture, Himachal Pradesh, took up the production programme in such a way so as to get more than two crops in one year from the cultivated areas. The introduction of improved technology in horticulture and agriculture in Himachal Pradesh has made it a glasshouse for the country in respect of seed potatoes, vegetable seeds, etc.

AGRO-CLIMATIC CONDITIONS

Nature has endowed Himachal Pradesh with a wide range of agro-climatic and topographical conditions. The Pradesh has subtropical climate in the lower and most of the valley areas, moderate in a few pockets, temperate in higher hills and dry climate prevails in areas like Kinnaur, Lahaul and Spiti and Pangi tehsil of Chamba district. The average annual precipitation ranges from 20 cms in the cold desert zone to 250 cms in Kangra valley, about 75 per cent of which is received from June to September.

Soils are almost neutral in most of the areas and acidic in the high rainfall region like Kangra valley. The soil texture varies from clayey loam in high rainfall areas to sandy loam in cold desert (Spiti area). Based on agro-climatic conditions, the state can be broadly divided into four major zones as under:

(a) Shiwalik Hill Zone

It consists of foothills and valley areas upto an elevation of about 800 metres above sea level, with subtropical area and about 35% of the geographical area and about 33% of the cultivated area of the state. The soils are mostly sandy loam in texture with scattered loamy patches. The area is highly erodable due to weak geological formation and scanty vegetation. The moisture retentive capacity is poor. The area receives about 150 cms of annual precipitation, 75% of which takes place during monsoon season. The main crops grown are wheat, maize, paddy, gram, sugar cane, mustard, potato, vegetables and barseem. Two crops in a year are obtained in this zone.

(b) Mid Hills

This zone extends from 800 metres to 1600 metres above sea level having mild temperate climate with annual precipitation of about 180 cms, 70% of which is received during monsoon season. The soils vary from sandy loam to loam in texture. It occupies about 32% of the total geographical area and about 53% of the cultivated area of the state. The area is relatively less subjected to moisture stress conditions. Major portion of the uncultivated area is under grasslands, mostly on southern and western slopes. The forests are mainly met with in northern and northern-eastern slopes.

The common field crops grown are maize, wheat, mash beans, barley, paddy, etc. In this zone two crops in a year are obtained. This zone is very potential for cultivation of crops like off-season vegetables and ginger.

(c) High Hill Zone

It lies above 1,600 metres above sea level with humid temperate climate and alpine pastures. The soils are mostly clayey loam to loam in texture with mild acidic reaction. This zone covers about 25% of the geographical and about 11% of the cultivated area of the state. The annual precipitation varies from 100 to 150 cms, 60% of which takes place during monsoon season. There are good pastures and meadows, which support rearing to a considerable extent. In the high hill zone mainly one crop in a year is obtained. But nowadays, in some areas, people grow pea as a cash crop after harvesting wheat and barley.

(d) Cold Dry Zone

It comprises Lahaul & Spiti and Kinnaur districts and Pangi tehsil of Chamba district lying about 2,700 metres above sea level. It occupies about 6% of the geographical and 3% of the total cultivated area of the state. The area is endowed with dry, temperate conditions having precipitation less than 20 cms during summer and very heavy precipitation in the form of snow during winter. Being a single crop area, the cultivation is carried out in summer season under irrigated conditions. The soil strata is very loose and is prone to heavy erosion due to glaciers and avalanches. It is ideally suited for the production of quality seed potato.

HORTICULTURE

Horticulture has played an important role in improving the economy of Himachal Pradesh. The typical geographical features coupled with favourable agro-climatic conditions have helped in changing the cropping pattern in the economy of farmers.

The area under fruits has increased from nearly 1,82,866 hectares in the year 1993-94 to 2,12,951 hectares in 1999-2000 and 1,91,668 in 2005-06. The area under fruits and total fruit production in 2005-06 is given in Tables 11.2 and 11.3. Favourable and varying agro-climatic conditions suit the development of subtropical, temperate and dry varieties of fruits in

Table 11.2 Area under fruits

Year	Apple	Citrus	Nuts & Dry fruits	Other fruits	Total (Hectares)
1	2	3	4	5	6
1993-94	72,406	37,961	14,553	57,946	1,82,866
1994-95	75,469	38,323	14,935	60,962	1,89,689
1995-96	78,292	38,595	15,237	63,560	1,95,684
1996-97	80,338	38,369	15,478	62,027	1,96,212
1997-98	83,056	38,635	15,832	64,839	2,02,362
1998-99	85,631	38,711	16,061	66,837	2,07,240
1999-00	88,673	39,138	16,396	68,744	2,12,951
2000-01	88,673	39,138	16,396	68,744	2,12,951
2001-02	92,820	40,174	16,956	73,085	2,23,035
2002-03	81,630	19,784	10,700	64,092	17,6206
2003-04	84,112	20,261	10,939	67,129	1,82,441
2004-05	86,202	20,402	11,100	69,199	1,86,903
2005-06	88,560	20,729	11,210	71,169	1,91,668

Source: Horticulture Department, Himachal Pradesh

Table 11.3 Production of fruits

Year	Apple	Citrus	Nuts & Dry fruits	Other fruits	Total (' 000 Tonnes)
1	2	3	4	5	6
1993-94	294.73	4.41	2.21	24.12	325.47
1994-95	122.78	6.67	2.37	38.72	170.54
1995-96	276.68	5.84	2.47	26.90	311.89
1996-97	288.54	13.83	3.35	45.91	351.63
1997-98	234.25	11.76	2.45	31.23	279.69
1998-99	393.65	13.11	3.07	37.85	447.68
1999-00	491.30	9.26	1.89	29.13	89.41
2000-01	376.73	11.06	2.75	37.49	428.03
2001-02	180.53	20.46	2.91	59.55	263.45
2002-03	34826	16.03	3.25	92.08	459.62
2003-04	459.49	28.12	3.57	68.79	559.97
2004-05	527.60	28.55	3.73	132.13	692.01
2005-06	540.36	29.16	3.92	122.08	695.52

Source – Horticulture Department, Himachal Pradesh

southern, middle, and the northern inner zones, respectively. In midzone, all localities are suitable for apple and temperate fruits except the sun backed slopes.

ANIMAL HUSBANDRY

Livestock is the backbone of the region's agrarian economy, which primarily rests upon manual and draught animal power and it also is a necessary adjunct to land and unremunerative nature of its cultivation. Animal husbandry is an undoubted source of additional income to the people living in higher hills and in mountain locked tribal areas. In the higher hills, sheep are like a poor man's cash crop and goat's subsistence, the former supplies him clothes and the latter milk and meat. The livestock population statistics as per the 2003 census are given in Table 11.4.

Table 11.4 Livestock census in Himachal Pradesh

	Category *Livestock and Polutry*	*1992*	*1997*	*2003*
1.	**Cattle**	21,65,034	20,01,826	21,96,538
2.	**Buffaloes**	7,03,549	6,52,373	7,73,229
3.	**Sheep**	10,78,940	9,088,831	9,06,027
4.	**Goats**	11,18,094	9,46,529	11,15,587
5.	**Horses & Ponies**	14,055	22,026	17,144
6.	**Other Livestock**			
	Mules	16,390	24,404	23,938
	Donkeys	7,522	6,639	8,859
	Camels	367	168	137
	Pigs	7,286	4,670	2,795
	Yaks	5,690	2,548	1,590
	Mithuns	6	731	200
	Rabbits	-	-	4,393
	Total Livestock	51,16,933	45,70,745	50,46,044
7.	**Poultry**	7,22,596	3,84,880	7,64,136
8.	**Dogs**	1,45,771	1,70,147	2,08,254

Source: Livestock Censuses - Directorate of Land Records & Animal Husbandry, Himachal Pradesh.

LAND USE PATTERN

The present land use pattern in the Pradesh has been given in Table 11.5. Around 38% of the total geographical area of the Pradesh, which amounts to 37,033 sq km has been classified as forests. Most of the forests belong to the government, yielding about 25 per cent of the revenue. Besides, they provide employment to nearly five lakh persons and grazing facilities to about 20 lakh animals. The forests also form the catchment area for big rivers like Sutluj, Beas, Ravi, Chenab and Yamuna.

Table 11.5 Land utilization in Himachal Pradesh (1999–2000)

Sr. No.	*Category of area*	*Area (sq km)*
i.	Geographical area	55,673
ii.	Forest area (Forest records)	37,033
iii.	Permanent pastures and other grazing lands including Alpine pastures, barren and incultivable wastes, etc.	7,529
iv.	Fallow lands (Current and other fallows)	719
v.	Net area sown	5,514
vi.	Cultivable wastes	1,194
vii.	Land under miscellaneous tree crops, etc.	642
viii.	Land put to non-agricultural uses	3,022

Source: Directorate, Land Records, Himachal Pradesh.

The background clearly gives an idea that in certain areas only one crop is grown. In some such areas, Morels are predominantly collected for compensating for the second crop need. The survey in the present studies was undertaken to investigate the impact of morel collections on the socio-economics of the people of morel producing areas and the observations are presented in the following pages.

PRODUCTION AND TRADE OF MORELS (FIG. 11.1)

An estimated 50-60 tonnes of dry morels are exported every year from India to the international markets. Himachal Pradesh, Jammu and Kashmir and Uttaranchal are the main exporters of morels in India. The production of morels during 1981-82 from Himachal Pradesh (Annual Report of Forest Department) was put at 11,000 kg, whereas in 1982-84, it has been put at 48,3000 kg. In 1986-87 it has been estimated to be 55,200 kg. Our survey in different parts of Himachal Pradesh substantiates and substantially increases these figures as the production of morels during 2001-2002 was 20,000 kg, whereas, in 2003-2004 it has been estimated to 58,000 kg and during 2005-2006 it has been estimated again very less as 20,000 kg. For instance, in Banjar valley alone, which constitutes just a small fraction of entire morel producing area, a single local shopkeeper gathered from people 4,000 kg morels in 2003. Similar collection figures have been obtained from other parts of the Pradesh. By these figures the total morel collection far exceeds the figure arrived at by the forest department. If Banjar valley is taken as an indicator, the total morel producing area can be divided into thirty- two such sectors and the total morel production would be expected at 80,000 kg. (Fig. 11.1)*. This seems to be a logical figure because all morels collected do not go into the records of the forest department. At present the forests are leased out to individual contractors by the forest department for morel collection, and morel collection by others is prohibited. But in practice, there is pilferage and many other sections of the society, e.g. villagers, labourers and school children also collect morels. The various collectors sell their fresh or dried collections

*Each *Morchella* fruiting represents a single district in the figure and each district has many morel producing valleys.

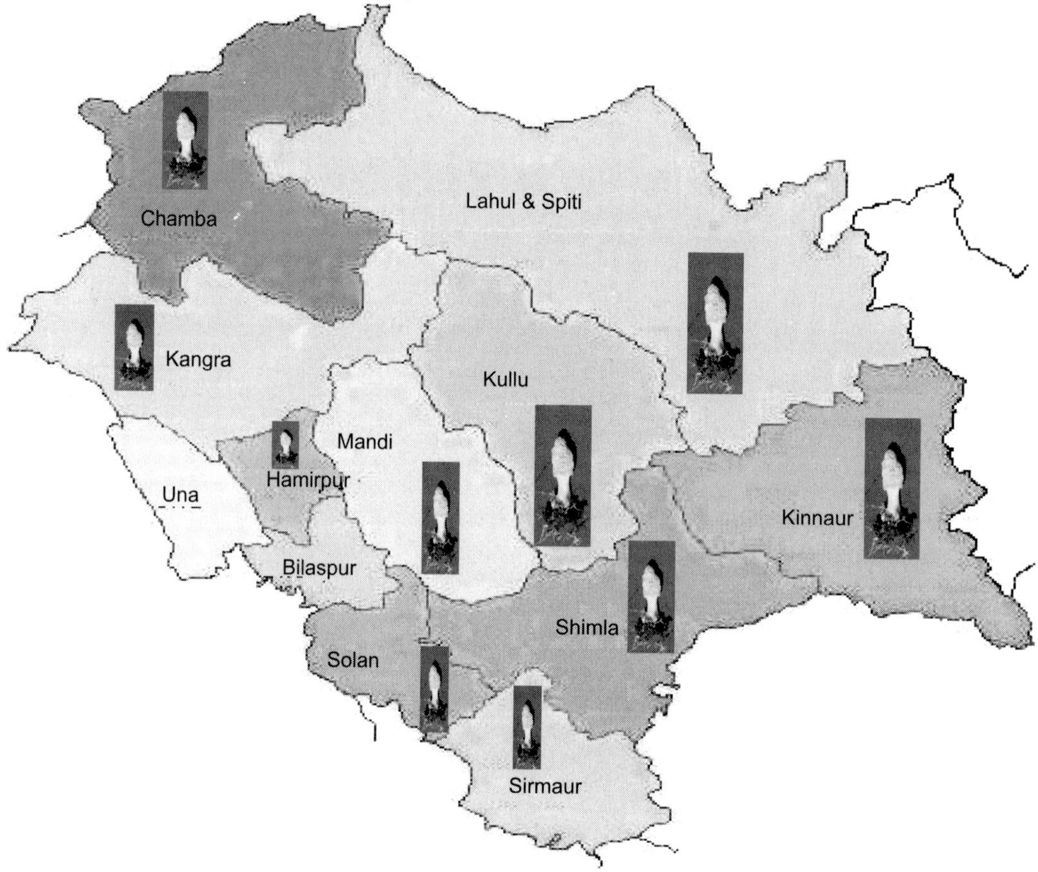

Fig. 11.1 District-wise presentation of *Morchella* producing valleys in Himachal Pradesh

at farm site at the rate of Rs. 1000-2000 per kg, to the wholesaler at the rate of Rs. 5000-60000 per kg and these ultimately reach the consumer at the rate of Rs. 8000-10,000 per kg. This shows that final consumer pays three-five times more price than the price at the farm site and real benefit doesn't go to the collector. This huge share of the middlemen can be avoided if the market is channelized by opening morel cooperatives or nationalization of morel crop. This gap in the recorded and unrecorded production can also be narrowed down or eliminated if proper licensing policy is adopted.

SOCIO-ECONOMIC PROFILE OF MOREL COLLECTING POPULATION

A sample survey was conducted by interviewing people from selected families in selected villages in the morel inhabiting zone, in order to know the social and economic status of morel collecting population. For this purpose a schedule was designed and got filled through personally interviewing morel collecting people in different areas (Annexure – VIII).

Analysis and Interpretation of Data

Table 11.6 shows that 42% families engaged in morel collection have income upto Rs. 500/- while families with annual income upto Rs. 1,000/- constitute 17% of the collecting group. Further, it may be seen that another 16% and 12% of collectors are parts of the group having annual income upto Rs. 5000/- and upto Rs. 10,000, respectively. Only 13% of these families are having annual income above Rs. 10,000. From the foregoing analysis it is clear that the percentage of morel collecting families is inversely related to income.

Table 11.6 Annual income (from all sources other than morel collection) and morel collection

Annual income (Rs.)	*Percentage of morel collecting families*
Upto 499	42
500-999	17
1000-4999	16
5000-9999	12
More than 10,000	13
Total	100.00

Table 11.7 shows that 43% of the families surveyed have members ranging from 6-10, 28% have less than 5 members and 29% have more than 10 members each. The table also shows that large sized families are able to spare some family members for morel collection. It appears that larger the family, the more is reliance on morel collection.

Table 11.7 Family size and morel collection

Family members	*Total families surveyed*
Less than 5	28
6-10 and above	43
More than 10	29
Total	100.00

Table 11.8 reveals that large majority of the families engaged in morel collection are poor having medium size family and less than 2 acres of land. Smaller the landholding, more is the probability of morel collection being a supplementary source of family income.

Table 11.8 Land area owned by family and morel collection

Land area (acre)	*Percentage of morel collecting families*
Upto 1	15
Upto 2	66
More than 2	19
Total	100.00

Table 11.9 shows that very few families in the 'very low income' group or 'low income group' consume morel and large number of families in high income groups are consuming morels. Morel consumption is significantly and positively related to income of the families.

Table 11.9 Morel consumption by the collecting families

Annual Income (Rs.)		*Number of Morel consuming families*	*Percentage of Morel consuming families*	*Percentage of Morel collecting families*
Upto	500	2	6.67	42
Upto	1000	2	6.67	17
Upto	5000	3	10.00	16
Upto	10000	10	33.33	12
More than	10000	13	43.33	13
Total		30	100.00	100.00

$X^2 = 61.04$ significant at .01 level of significance.

Table 11.10 reveals that morel collection adds Rs. 500 to 5000 to the income of the poor people. Out of total families surveyed, 20 per cent got 500-1000, 32% Rs. 1000-2000, 18% Rs. 2000-3000, 16% Rs. 3000-4000 and 14% Rs. 4000-5000 additional income. Thus, *Morchella* is a source of additional income for the marginal farmers in the high hill zones of the Pradesh.

Table 11.10 Income from morel collection

Additional income (Rs.)	*Percentage of Morel collecting families*
500-1000	20
1000-2000	32
2000-3000	18
3000-4000	16
4000-5000	14
Total	100.00

The second part of the schedule provided information about the licence holding for morel collection and marketing of the collections. Hardly any member of high income families was engaged in morel collection. As many as 80 per cent of the families were not holding any licence and only 20 per cent families were holding licence for morel collection. This shows utter failure of the machinery to take care of morel collection on commercial basis or protection of morel collecting families from the brokers in the market who purchase morel at very low rate to sell on much higher prices in the market.

In nutshell, the poor peasantry in Himachal hills uses morel collection as a source of supplementing their income. The restriction on the consumption of morels and the high prices this commodity commands deters the local population from using morels for self-consumption except on some festive occasions. Thus, if morel collection is promoted, a sizeable surplus of this commodity shall be available for marketing in the domestic markets as well as for exports to the international markets.

CHAPTER

12 Mycorrhizal Association of Morels

CHAPTER OUTLINE

- Introduction and Review of Literature
- Observations and Discussion

INTRODUCTION AND REVIEW OF LITERATURE

Morels are a cherished delicacy and an important minor forest produce with enormous economic potential. Various attempts to domesticate them have been made since very early times but with little success. The indoor artificial cultivation of mushrooms has been successfully achieved in about fifteen species all over the world. This has, however, been possible only in the case of non-mycorrhizal mushrooms, where a successful imitation of natural process has been possible.

The morels were not known to enter into any mycorrhizal relationship till we briefly reported this association in *Morchella esculenta.* Later a similar but more detailed and definitive relationship between *M. rotunda* (Pers.). Boudier and some living trees and herbaceous plants was described and discussed by Buscot and Roux (1987).

Buscot and Roux (1987) reviewed the literature and the association of morels. According to them the morels have been generally believed to be parasitic, because they have been frequently observed to be associated with roots of plants in Oleaceae or Cornaceae (Robert, 1865). Similar associations between *Morchella esculenta* (L) Pers. and tubers of *Helianthus tuberosus* L by Roze, (1882 and 1883: cited as Buscot and Roux, 1987) have also been reported. More recently, the genus *Morchella* has generally been regarded as saprophytic (Heim, 1936 and Delmas, 1974) even though associations were reported between the psammophile *Morchella spongiola* Boud. and roots of a graminaceus plant (see Buscot and Roux, 1987). The evidence of saprotrophy is largely based on experiments showing that morel mycelium can grow on a wide range of artificial culture media (Brock, 1951; Robbins and Hervey, 1965; Litchfield, 1967 and Kaul, 1977) and by the occasional fructifications produced on such media (Molliard, 1905; Cailleux, 1969 and Ower, 1982).

Morchella esculenta, M. conica, M. angusticeps and *M. deliciosa* have been observed .o enter into mycorrhizal association with herbaceous plants, i.e. strawberry, fern and grasses. The subterranean portions of the carpophore form a loosely woven cord near its base. Morel hyphae penetrate into all the tissues of roots except xylem. The hyphae form a loose weft on the root periphery, some of them penetrating directly into the cell lumens (Lakhanpal *et al.*, 1991) (Table 12.1).

Table 12.1 Mycorrhizal association between different *Morchella* species and host plants

S.No.	*Morchella spp.*	*Host plant*	*Reference*
i.	***M. rotunda***	Members of families Oleaceae and Cornaceae	Robert (1865)
ii.	***M. esculenta***	Tuber of *Helianthus tuberosus*	Roze (1882, 1883 cited as: Buscot and Roux, 1987)
iii.	***M. spongiola***	Graminaceous plants	Romagnesi (1963)
iv.	***M. deliciosa***	Several grasses	Lakhanpal (1986)
	M. crassipes	Several grasses	Lakhanpal *et al.* (1991)
	M. esculenta	Fern rhizomes	Lakhanpal *et al.* (1991)
	M. conica	Raspberry	Lakhanpal *et al.* (1991)
v.	***M. rotunda***	*Ligustrium vulgare, Cornus sanguinea*	Buscot and Roux (1987)

A more detailed investigation of such a type of association of *M. rotunda* was published by Buscot and Roux (1987). In their report they observed that ascocarps of *Morchella rotunda* are joined by subterranean hyphal systems to "mycelial tuffs" surrounding living roots of various plants. They reported that from the "mycelial muffs" the fungus penetrates into the root tissues as far as the vascular cambium and grows predominantly intercellularly, especially within secondary phloem. They noted that formation of this close association with plant roots may represent a critical stage, concerned with elaboration of the ascocarps in the life cycle of this otherwise saprophytic fungus. They suggested that it would be necessary to investigate whether other species of *Morchella* also form mycorrhizae or not.

In the meantime, we observed other morel species to enter into mycorrhizal relationship. The observations on the same relationship are being presented here.

Mycorrhizal connections were established by digging the soil and tracing the roots of the plant species to the fructifications following Young (1936, 1940) and Zak (1971, 1973). The roots of plants associated were thoroughly washed with tap water, and after fixing them in FAA for 24 hours they were preserved in 70% alcohol. In order to confirm the formation of mycorrhizae, both stained (Cotton blue in Lactophenol) and unstained sections were prepared and examined microscopically (Annexure – IX).

OBSERVATIONS AND DISCUSSION

Observations for mycorrhizal association between the sporocarps of *Morchella deliciosa*, *M. esculenta* and *M. conica* and the roots of strawberry, grasses and fern rhizomes were made visually. The sporocarps were taken out carefully along with some mass of soil. These were allowed to dry and then freed from soil particles or they were washed slowly in running water near its base, which was gradually transformed into a more compact and shapely cord, producing some short roots on way, and tapering at the distal end and there getting connected to the farther end of the roots of these plants, forming a 'mycorrhizal bridge' (Fig. 12.1 a–c; Plate I a–f).

Anatomical observations made on sections stained with cotton blue show that morel hyphae penetrate into all the tissues of roots except xylem (Fig. 12.1 d). The hyphae penetrate into the cells where they grow mostly near the cell wall, but they also sometimes grow more deeply within the lumen. The root hairs are mostly absent. The typical fungus mantle was also absent. The hyphae have not been seen to grow intercellularly. In long sections of the root, the hypae were seen to form a loose weft on the root periphery; some of them penetrating directly into the cell lumens (Fig. 12.1 e). The short roots also show a similar structure and they are almost evenly distributed all along the long root and its branches.

As has been observed in the present investigations, Buscot and Roux (1987) also had noted that ascocarps of *Morchells rotunda* are joined by subterranean hyphal systems, the 'mycelial muffs', surrounding living roots of various plants. Anatomically, the fungus has been found to penetrate into the root tissues as far as the vascular cambium, and grows predominantly intracellularly, especially within young secondary phloem. The formation of this close association with plant roots may represent a critical stage, concerned with elaboration of the ascocarps in the life cycle of this otherwise 'putatively saprophytic fungus.' They mention that the localization of 'muffs' on parts of roots that are non-absorptive,

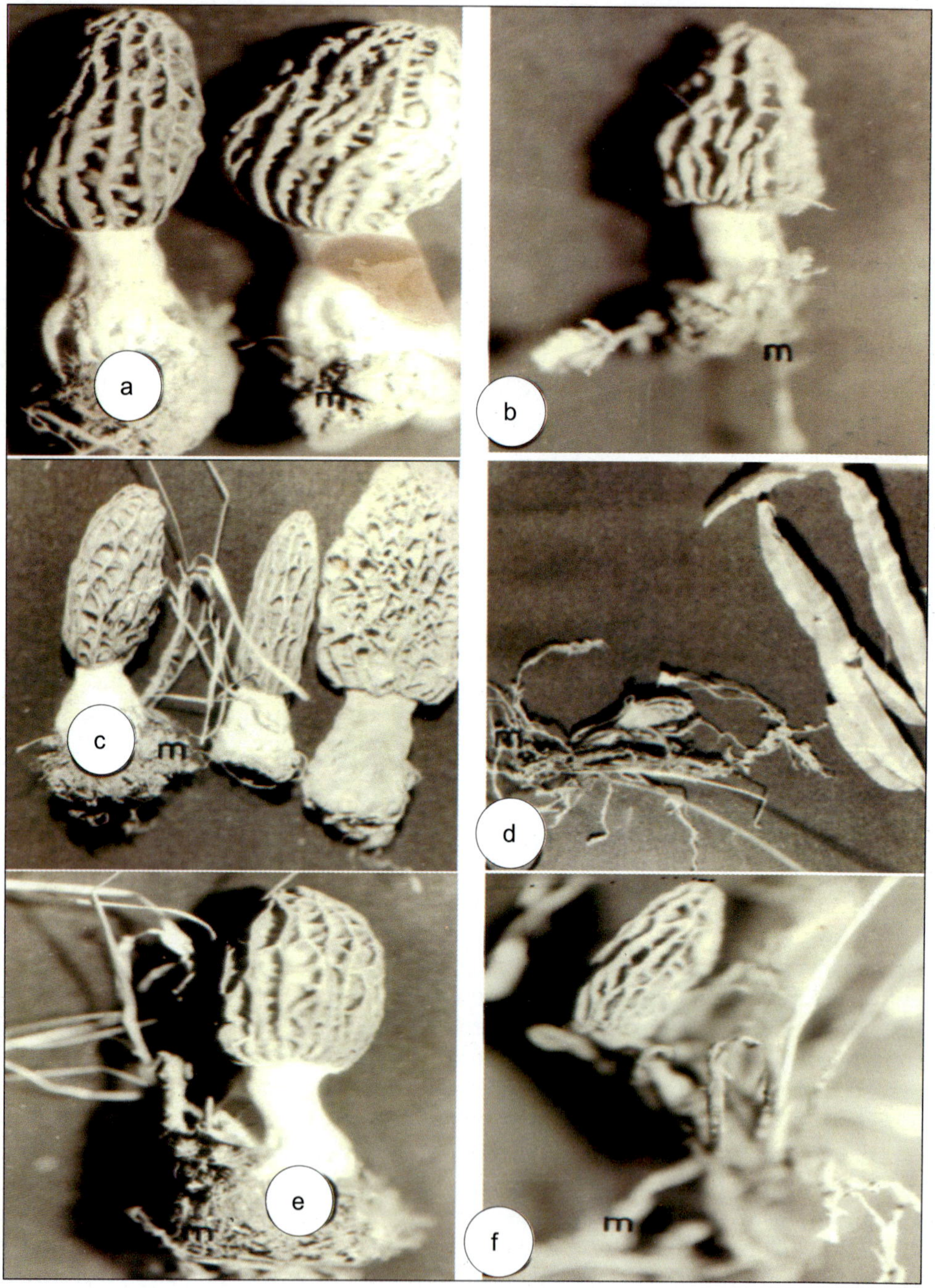

PLATE I (a–f) *a. Morchella crassipes* showing mycelial cords with grass roots. *b. Morchella conica* showing mycelial cords with grass. *c. Morchella crassipes* in mycorrhizal association with grass roots. *d. Morchella deliciosa* in mycorrhizal association with fern. *e. Morchella esculenta* in mycorrhizal association with grass. f. *Morchella esculenta* in mycorrhizal association with fern.

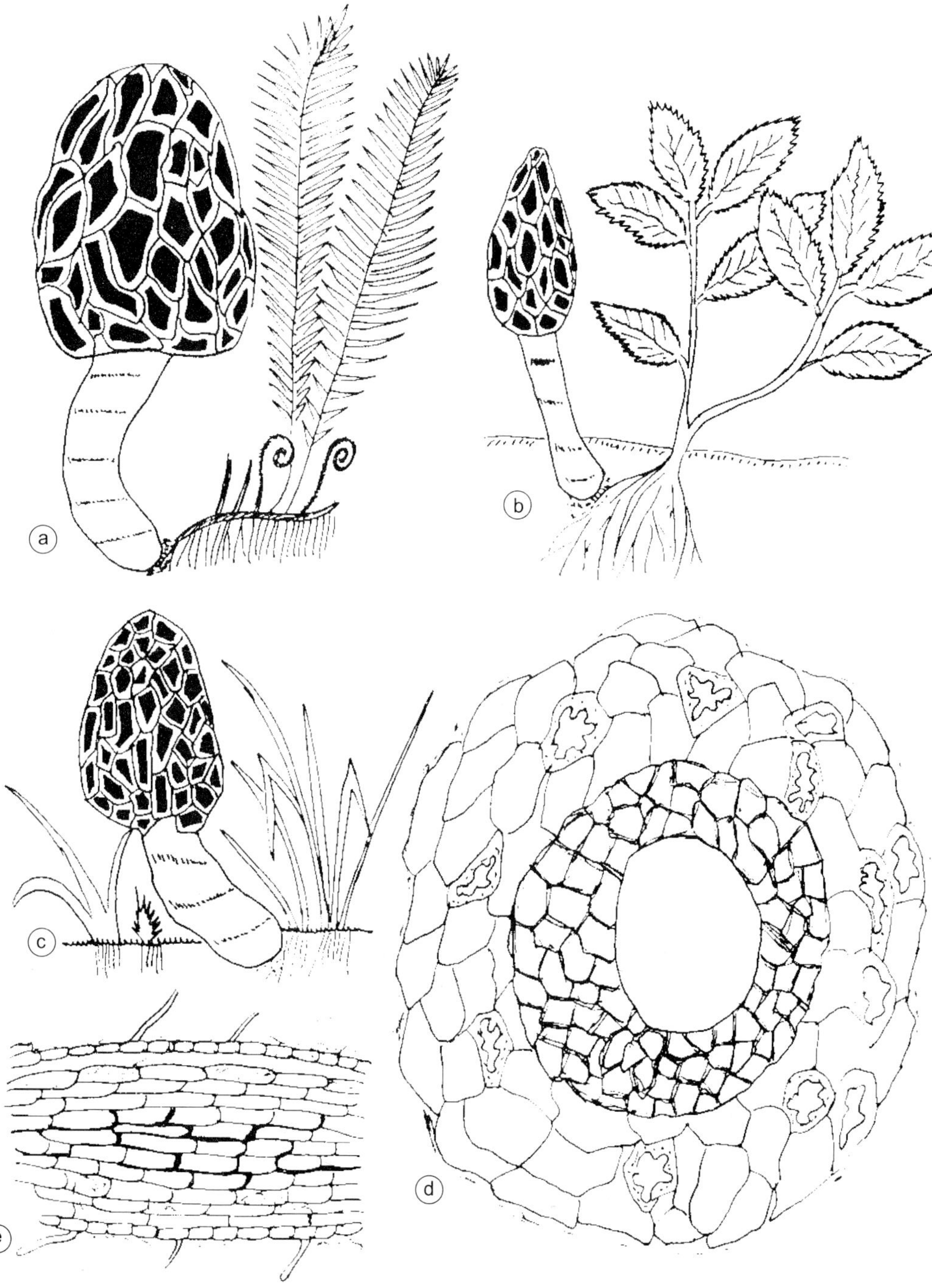

Fig. 12.1

suggests that the association that we have observed between species of morels and the herbaceous plants during the present studies seesms to be doubtlessly mycorrhizal, as the fungus has been found to be associated with young absorbing roots.

Buscot and Roux (1987) also suggest that this morel may possess a parasitic or symbiotic phase in its life cycle, and that consistent with this possibility is the direct relationship between the ascocarp and the 'mycelial muffs' exclusively located on subterranean organ that conduct elaborated metabolites and the preferential arrack of tissues within that are concerned with their transport. They further pointed out that on the other hand, the limited area colonized by the fungus and the saprophytic potential of its mycelium suggests that either this parasitism is temporary or that the association was more to do with stimulation of ascocarp formation than nutrition per se.

Though all these morel species were collected from forested areas with many species of conifers predominating, their association with trees could not be established. Circumstantial evidence does suggest that some species may form mycorrhizae with apple trees. The fruiting bodies of morels have been frequently observed to crop up in the apple orchards, especially around the apple tree basins. In the forests where the fruiting bodies come up, they apparently do not seem to form any mycorrhiza with them. However, considering reports of Schramm (1966), and Maroneck *et al.* (1981) that the mycorrhizal strands of *Pisolithus tinctorius* may extend 5-10 m from the seedlings to the basidiocarps, the association of morels also needs to be examined more closely and carefully to negate any such link. If it was purely an association with herbaceous flora, then these should be expected to come up in areas which do not have trees but abundant herbaceous flora. We have observed them to come up in pastures, but not in that great abundance.

The mushrooms that enter into symbiotic mycorrhizal relationship with different plants are usually host specific and hence cannot be grown indoors. In such cases, fields cultivation has been suggested and practised in some instances. Attempts to obtain fruiting bodies of mycorrhizal mushrooms without host plants have never been successful because such mushrooms accomplish their biological cycle jointly with the host plant, giving rise to a complex nutritional relationship.

The observations that morels also enter into mycorrhizal association are of utmost significance. These are a pointer to the fact that may be it is the mycorrhizal associations, which has so far foiled all attempts at artificial cultivation of morels. It seems now imperative to direct attention on this aspect, and if it is so, to alternatively try their cultivation outdoors with specific host plants.

CHAPTER 13

Cytology of Ascus Development in Morels

CHAPTER OUTLINE

INTRODUCTION

The general life cycle pattern in ascomycetes is almost the same in the class as a whole with some deviations in simpler groups. The most often cited fungus which typifies this life cycle is the species of *Pyronema*. These species were initially worked out very early (Claussen, 1912) and have also been studied in the recent years with improved techniques, i.e. Scanning Electron Microscopy (SEM) and Transmission Electron Microscopy (TEM) (Hung and Wells, 1971), confirming the earlier observations. Cytological studies on fungi have lagged far behind than other groups of organisms. This has been primarily and the frequent inability of the latter to take the more usual chromosomal stains (Burnett, 1976). However, some of the hurdles have been overcome with the use of electron microscopy. Most of the studies on life cycle and ascus development have been done on organisms, which can be easily grown in culture. Many of the ascomycetes produce definite sex organs, whereas in others there are no recognizable sex organs and fusion in them takes place between somatic hyphae. The crozier formation in some is stated to be typical, while in others, especially the higher ascomycetes, it is said to be absent. Even with regard to *Morchella* contradictory views have been expressed in respect of crozier formation and ascus development (Maire, 1905; Gries, 1940).

REVIEW OF LITERATURE

Studies on the cytology of the fungi were started in the beginning of this century. Harper (1897, 1905) was the first to study the meiosis of powdery mildews. Later after four decades, McClintock's (1945) studied asci of *Neurospora*. Singleton (1953), while working on *Neurospora crassa* confirmed the findings of McClintock (1945). Singleton (1953) observed that the conjugate premeiotic nuclear divisions in the croziers are meiotic, with set of chromosomes (7 in *Neurospora)* passing to each spindle pole. Fincham (1949) established that 7 is the common haploid number throughout the genus *Neurospora.* St. Lawrence (1952) identified for the first time specific chromosomes in fungus with known linkage groups using X-ray mutants in *N. crassa.* carrying translocations.

Singer (1961) observed that in *Morchella* spp. sometimes a diakaryotic hypha directly develops into an ascus and there is minimum development of ascogenous hyphae. Ower (1982) reported that the fruiting body develops from a single cell, while Gessner *et al.* (1987) claimed that it originates from a conglomerate of sterile haploid tissue. The ascus mother cell is multinucleated and no hook cells or croziers have been observed. The two paired nuclei migrate to the tip of the ascus mother cell. These fuse to form a large diploid nucleus with the large vacuole behind. Meiosis occurs followed by four successive mitotic divisions and the ascospores are delimited by free cell formation. The mature ascus contains eight ascospores. Each ascospore is multinucleate with eight nuclei. The ascus is operculate and eight ascospores are released explosively, simultaneously and may travel upto several metres (Schmidt, 1979) and leave the surface of fertile portion of carpophore like smoke clouds.

Cytology of the vegetative mycelium of the genus *Morchella* was studied by Gries (1940) and Berhet (1964). They reported that hyphae are septate, with several nuclei per cell. Gries (1940) also reported that plasmogamy occurs very late in the development of the ascorcarp, i.e. in the subhymenial layers just before the formation of asci. According to him morphologically undifferentiated cells in these layers fuse, and from them arises a short ascogenous hypha containing paired nuclei, but it is not clear whether only one ascus arises from each of these hyphal fusions. The ascus develops from the tip cell of this ascogenous hypha without formation of croziers. However, earlier, Maire (1905) had reported that the cytology of ascus development is normal and eight ascospores are formed. Korf (1973) reported that in the family Morchellaceae the ascopores are always multinucleate and have large number of tiny, external guttules in the epiplasm, at both ends. Therefore, it appears from all these studies that the pattern of ascus development is still unclear.

The standard methodology followed for fixation and staining is given in Annexure -X. In the following pages are presented observations made on the ascus development of different species of *Morchella* collected from North-West Himalaya.

COLLECTION AND FIXATION OF MATERIAL

The *Morchella* sporocarp is a composite apothecium. It was initially thought that each pit or depression in it would represent an individual apothecium, and since one sporocarp has small as well as large depressions, they should possess ascogenous hyphae and asci at different stages of development. However, it was observed that immature fruiting bodies, even the smallest ones, possessed only mature or partially mature ascogenous hyphae and asci. So, a search was initiated to collect sporocarp primordia of different *Morchella* species and fix them for cytological/developmental studies. In such primordia divisional and developmental stages of the desired structures could be observed. But at primordial stage all the species were indistinguishable morphologically. So, the primordial stages were fixed from sites where mature fruiting bodies of particular species were also present. In this way, it was possible to fix developmental stages of five species only. *M. semilibera* could not be collected at primodial stage. Since different species were distributed in different forest zones and were found associated with different trees, primordia were also collected from such different zones and sites and worked out for developmental stages. The preparation did not come up very well. Since some interesting stages were recorded, it was considered worthwhile to report them.

OBSERVATION AND DISCUSSION

The stages in the development of the crozier and karyogamy in *M. esculenta, M. conica, M. deliciosa, M. angusticeps* and *M. crassipes* are presented in Fig. 13.1 and Plate I (a–f) and Plate II (a–g), which were similar in all these species. Development of the crozier from the binucleate ascogenous hypha is typical of ascomycetes. The binucleate ascogenous hypha elongates and bends in the form of a hook (Fig. 13.1, a, b). Septa are laid down so that the ultimate or the tip cell has one nucleus, the penultimate cell has two nuclei, and the anti-penultimate cell has two nuclei (Fig. 13.1c). The ultimate and penultimate cells may fuse, and later other croziers may arise. This process results in clusters of croziers. The two nuclei in the penultimate cell fuse, forming ascus mother cell containing diploid nucleus surrounded by densely staining material (Fig. 13.1d). Soon after fusion, the young ascus with its diploid nucleus begins to elongate vertically and expands transversely (Fig. 13.1h). The fused nucleus soon undergoes

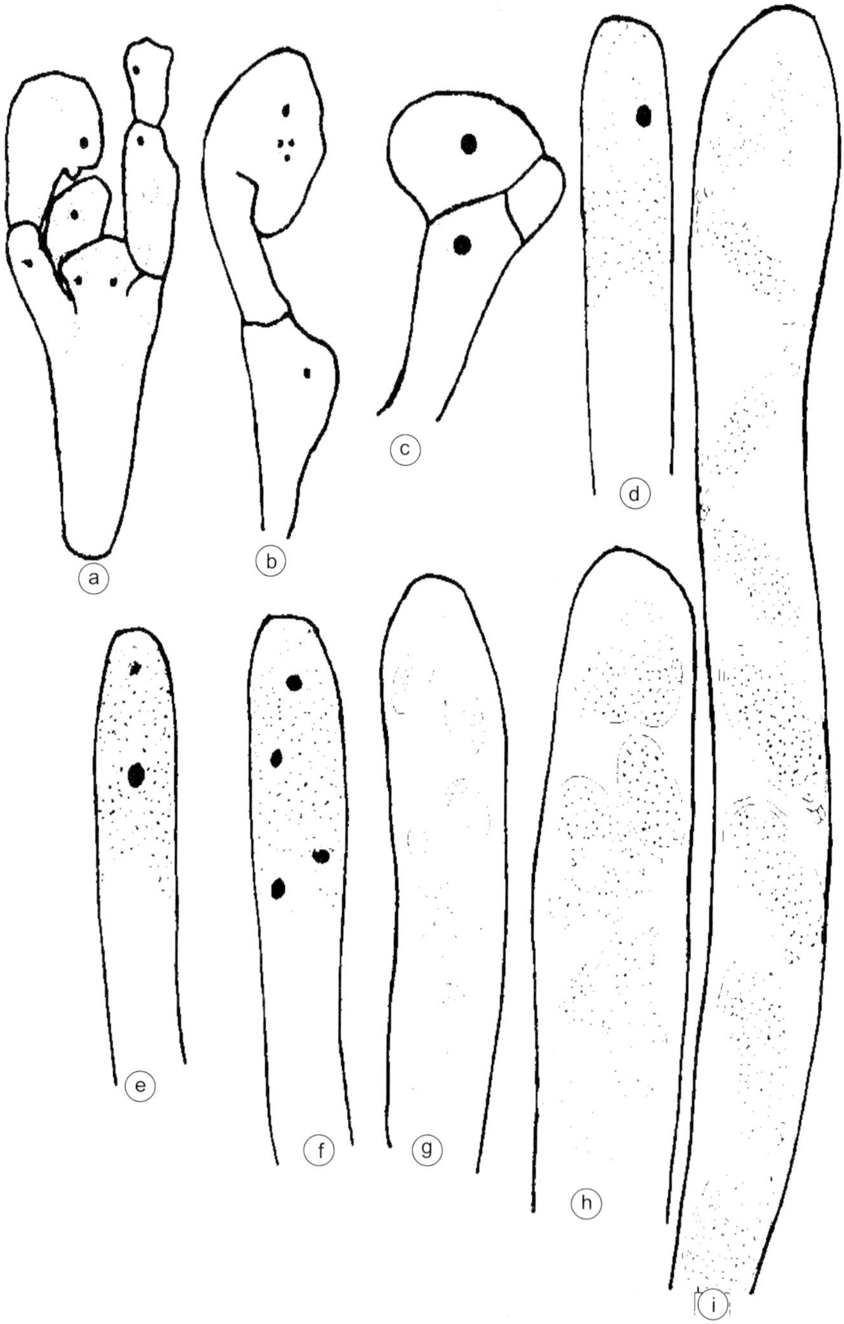

Fig. 13.1 (a–i) Ascus development in *Morchella* spp. (a) Ascogenous hyphae (b, c) Crozier formation (d) Ascus mother cell (e) Ascus with two nuclei (f) Ascus with four nuclei (g) Immature ascus with developing ascospores (h) Immature ascus (i) Fully developed ascus with eight ascospores

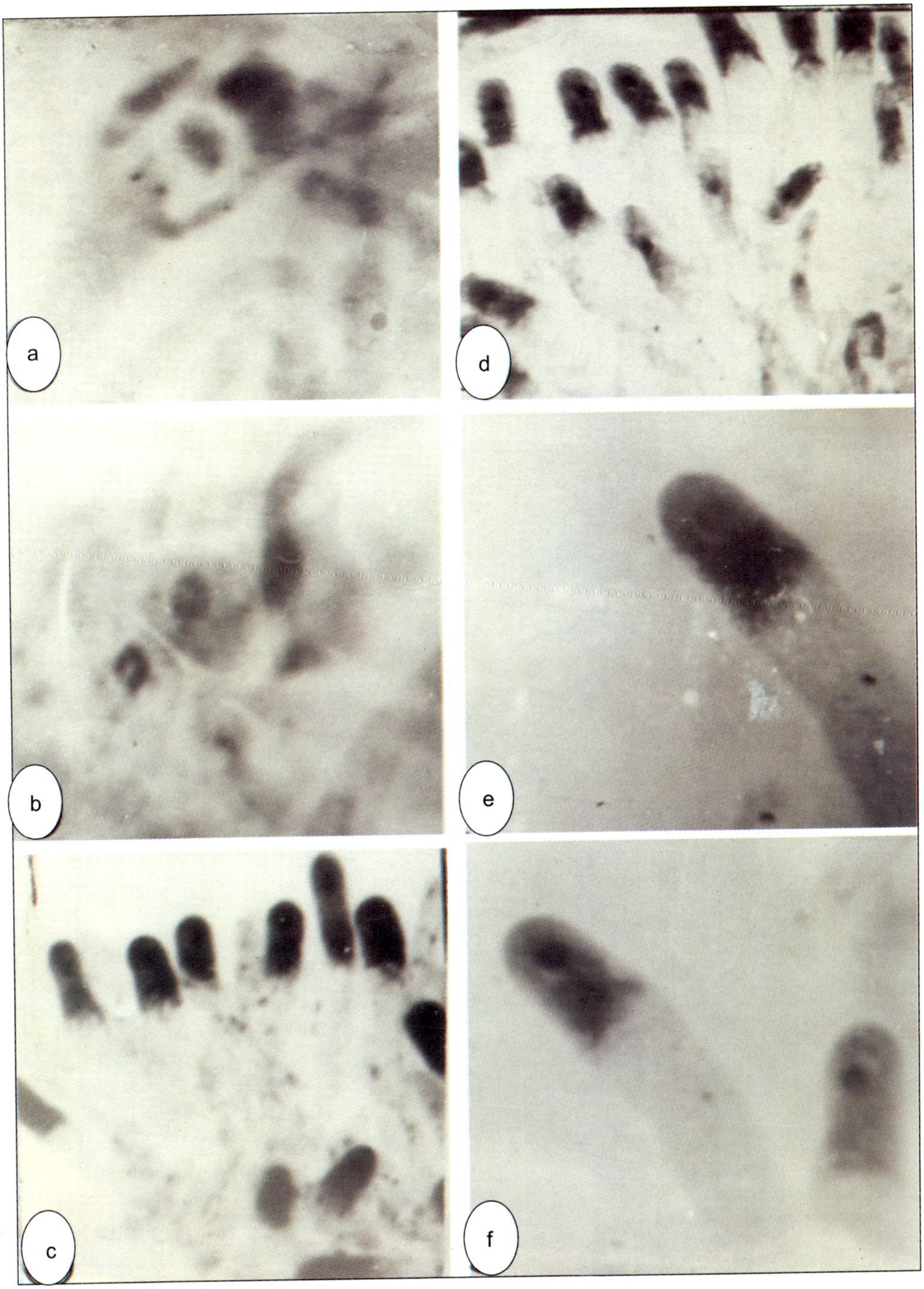

PLATE I (a–f) (a) Initial stage of crozier formation in an Ascogenous hypha. (b) Developing with a typical crozier u – ultimate cell, p – penultimate cell, a – anti-penultimate cell (c) Ascus mother cell, each with a single diploid nucleus. (d) Same at a slightly later stage, bur again each with a single distinct nucleus. (e) Enlarged ascus mother cell with a single nucleus (f) Ascus mother cell nucleus dividing into two (↓) and then initial four nuclei (↓).

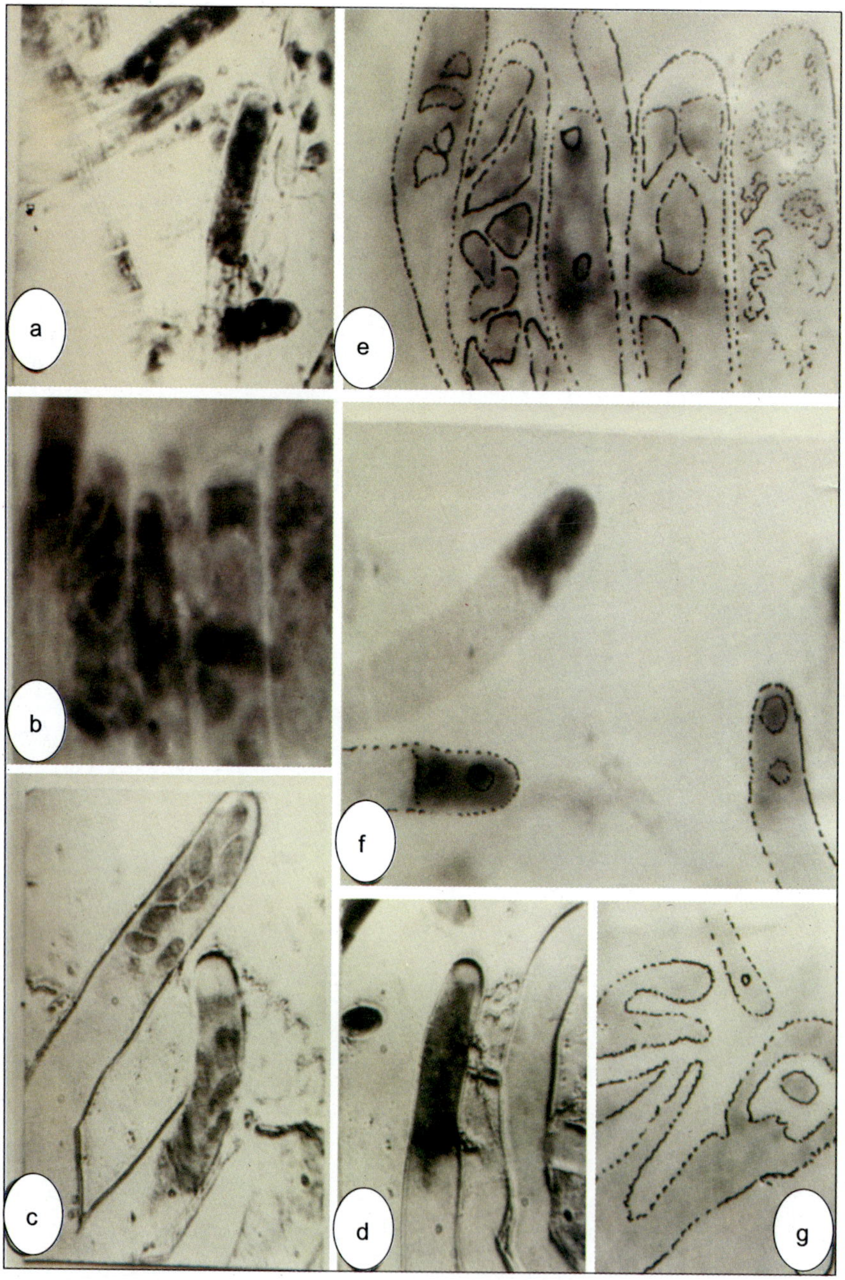

PLATE II (a–g) (a) Ascogenous delimitation within the developing ascus. (b) Ascus with eight immature ascospores. (c) Ascus (?) after the release of ascospores (↓). (d) h – photograph retouched to show delimitation of ascospores. (e, f) photograph retouched to show two nucleate stage. (g) 'a' – photograph retouched to show crozier formation.

meiosis, which results in four haploid nuclei (Fig. 13.1f), each one of them divides mitotically forming ultimately eight nuclei (Fig. 13.1g). These become incorporated into the ascospores, which develop by free cell formation (Fig. 13.1h, i). The ascospore nucleus divides mitotically and the mature ascospores are multinucleate. Paraphyses, which were seen arising between the asci, are multinucleate.

Gries (1940) reported that there is no crozier formation in *Morchella* spp. However, the crozier formation was clearly observed in *M. esculenta, M. conica, M. deliciosa, M. crassipes* and *M. angusticeps,* and hence in all the five species, the ascus development was typical of the other ascomycetes. Eight ascospores per ascus were formed. These studies confirm the observations of Mair (1905) that the cytology of ascus development in morels is normal and eight ascospores are formed.

The technique used in the present investigation, i.e. BAC fixative and Propiono-Iron Haematoxylin stain with slight modification in the staining procedure, gives the best preparation. The nuclei and the chromatin material stain deeply, while cytoplasm remains particularly colourless.

Nuclear behaviour of the ascus of *Morchella* species was quite similar in its main features to that of many investigated members of the ascomycetes (Singleton, 1953; Roger 1965; Lu, 1967). Similarly, karyogamy observed in the penultimate cell was similar to that reported by Olive (1965) for other ascomycetes. The ascospores were observed to be multinucleate in all the five species, confirming the findings of Korf (1973). The present investigations reveal that there was no difference between the various developmental stages of *M. esculenta, M. conica, M. angusticeps, M. deliciosa* and *M. crassipes.*

CHAPTER 14

Culinary Aspects of Morels

CHAPTER OUTLINE

- Introduction
- Review of Literature
- Observations

INTRODUCTION

The mushrooms have been consumed as food by man since very ancient times. They have been collected and consumed traditionally since antiquity. The mushrooms were traditionally collected from nature along with fruits and berries for consumption. The reason for their consumption was the belief that they were very good energy source. The Greeks regarded them to provide strength to their soldiers in wars, while Romans considered mushrooms as 'cibus diorum' as 'Food for the Gods' and Chinese as 'Elixir of Life' (Chang and Miles, 1982). The folklore of many cultures, including Russia, China and Mexico held that eating mushrooms could give some super human strength (Mateljan, 2005). In ancient Rome, meals which included mushrooms, were highly appreciated (Dermek, 1982).

However, all people were not amenable to mushrooms. In different cultures some feared them whereas others liked them. Hence, two types of cultures developed with regard to mushrooms: the Mycophilic and the Mycophobic. In the mycophobic or fungiphobic cultures, the mushrooms were feared and considered to have no food value, while in mycophilic, the mushrooms have been looked upon as staple food by some and gourmet's delicacy by others (Miller, 1981). In fact, the edible fungi have been considered useful and a table delicacy primarily because of their culinary properties and gastronomical delights and only secondarily for their nutritional and medicinal values. It is only the peculiar aroma, texture, flavour and taste for which the mankind has long devoured the mushrooms. A scientific approach for their nutritional and medicinal value is only a recent phenomenon.

Edible mushrooms are a group of extremely versatile organisms. They may be eaten in a variety of ways. They may be eaten raw or cooked in the form of gravies, as garnish to meat and vegetables, fried, fritters, baked, puddings, stuffed, sautéed, marinated, may be pickled, made into ketchups, croquettes, soups, creamed or used as major ingredient for omelettes, casseroles and snacks, etc. For every mushroom there is a right recipe, and if a right recipe for a right mushroom is used, the resulting culinary delight may be heavenly. The mushrooms may be dried or powdered and it is in this form that they are being used for seasoning throughout the world. The dried, powdered, canned or frozen mushrooms can also be stored for longer periods and used when they are not in season (Miller, 1981).

With the increased interest in cultivation of mushrooms and awareness about those occurring in the wild, there is a lot of development in consumption patterns and many recipes are available for satisfying gastronomical delights. The cultivation technology has made mushrooms the 'Food of Common Man' (Das, 2003), and they are served on all occasions and have become a part of menu for all festive occasions.

Among the wild mushrooms, the four major varieties are the cepe, chanterelles, morels and field mushrooms. Among all the wild edible mushrooms, morels are the most desired and relished ones. The genus *Morchella* is the safest to eat, and is unheard of causing even slightest uneasiness to the patron. The morels constitute the most prized and highly detectable table delicacies among edible mushrooms, with good protein value. They have intense and earthy flavour. The morels have been collected and eaten in India since times immemorial. In the Himalayan region, there is a mixed population of both tribal and non-tribal people. In many parts of Himalaya, the old traditions are still preserved. So, an attempt was made to record the methodology followed by the ethnic people as well as by modern society for the preparation of traditional and modern dishes of morels, setting forth the following objectives:

(i) To document the ethnic knowledge base for the consumption of morel species.

(ii) To record data on traditional recipes for the preparation of morel dishes.

(iii) To record changes in recipes which have emerged during the course of use.

The details of standard methodology for culinary aspects are given in Annexure -VII.

REVIEW OF LITERATURE

There are about 2,000 species of edible mushrooms known all over the world, and out of those almost two dozen are widely accepted for human consumption (Chang and Miles, 1982). From India, around 200 species belonging to 70 genera have been reported to be edible (Chadha, 1992). Throughout the history of mankind, fungi have been regarded with fear and fascination; sometimes revered, sometimes hated, but always considered mysterious. They have been a source of food since times of antiquity and there are many recipes for cooking fungi in a book written by Caelius Apicus in the third century A.D.; probably the oldest book of cookery in Europe (Singh, 1999).

During the survey from Himachal Himalaya, out of 70 edible species recorded, only 20 species have been observed to be regularly and traditionally consumed by the local inhabitants without any causality or fatality. Some such fungi are species of *Morchella, Helvella, Tuber, Hericium, Sparassis, Hydnum, Boletus, Trappeinda, Clavaria, Ramaria,* and *Albatrellus* among the non-gilled members and the species of *Lactarius, Russula, Cantharellus, Coprinus, Lepiota, Amanita* and *Lentinus* among the gilled fungi (Lakhanpal, 1994). Mushrooms are usually consumed fresh during the season of their appearance. Even among these the most preferred species are of *Sparassis, Lactarius, Cantharellus, Helvella, Morchella, Hydnum, Boletus, Clavaria* and *Ramaria* (Lakhanpal, 1994). These are not only consumed fresh but some of these are dried and preserved for consumption in future.

Miller (1981) listed many precautionary measures to be observed while collecting and consuming edible mushrooms, such as packing of each species separately after thorough cleaning, storing the packets in a cool place in hot weather before consumption, establishing the correct identity of the species, cutting each specimen into halves before cooking, to see the presence or absence of worms, putting fruiting bodies in cold water for sometime to drive out the worms. He also enlisted some specific precautions for specific varieties of mushrooms, e.g. in some polypores only specific portions of the fruiting body can be eaten, as in *Polyporus sulphureus* only outer margin is cut and used when young because rest of the fungus is tough to be eaten. While in case of coral fungi, a soft bristle brush is used to clean the fungi. In case of puffballs, the fruiting body is cut in half down through the point of attachment and the body is freed of worms and worm-holes. For *Coprinus comatus,* he advised that the fruiting bodies should not be gathered and stored in the refrigerator. They must be cooked immediately; if not, the enzymes of the fungus will disintegrate them.

For the preparation of morel dishes, Miller (1981) suggested that the fruiting bodies should be cut in halves down through the point of attachment because the interior in them is hollow; insects and occasionally slugs may often be found inside. If there are no worm-holes, the entire morels may be eaten after the inside is carefully washed. He pointed out that shiitake, matsutake, dried boletes, chanterelles, truffles from Europe, and dried *Auricularia* from China are used as seasoning agent in dry and powdered form.

Many precautionary measures for collecting and consuming wild edible mushrooms from North America have also been reported by Lincoff (1981). According to him never eat wild raw fungi, eat a very small amount at first and observe its allergic reactions, and keep fungi refrigerated until eaten. Lincoff (i.c.) suggested that beginners should first reserve for the dinner table those mushrooms, which are easily identifiable, and in such cases non-gilled fungi fall in this category of easily identifiable ones.

Dermek (1982) described the role of mushrooms in diet and observed that mushroom dishes are becoming increasingly popular these days as scientific knowledge about fungi offers everyone a new look at this food. He observed that mushrooms are successfully used in slimming, because they produce a feeling of fullness and they keep the digestive system busy. He also suggested certain guidelines for the preparation of dishes as: prepare only fresh and sound fruiting bodies of mushrooms and cook them carefully. Mushrooms are harmless when reheated provided they were put in refrigerator soon after they were cooked. But he does not recommend to keep the food containing mushrooms, as they are best eaten soon after being cooked. According to him to children under 10 years age, mushroom dishes are to be served in small amounts. He suggested that most common methods for the preservation of mushrooms are drying and pickling. He illustrated 24 mushroom recipes and 2 other recipes for pickling mushroom solutions.

Different types of mushrooms have different fragrances, flavours, and textures, as the taste of the people differs (Arora, 1991). Arora (i.c.) pointed out that some species like chanterelles are delicate and temperamental, therefore, demand special treatment and other robust and good species don't require any treatment. His emphasis was on simple and elegant dishes of mushrooms that highlight flavour and texture of the mushrooms rather than drowning in butter, etc. He provided some guidelines for the preparation of the recipes and recommended two best methods of preservation: drying and freezing. According to him after reconstituting them some don't revive, but morel species revive very nicely after rehydration by regaining their original taste and texture.

For mushrooms, which are cultivated, a variety of recipes have been promoted. Thakur (1990) suggested that in India three types of cultivated mushrooms, viz. white button (*Agaricus bisporus*), dhingri (*Pleurotus sajor-caju*) and paddy straw (*Volvariella volvacea*) have been introduced as vegetables in the Indian menu and gave the recipes on Indian blend. She emphasized on the fact that people have diverse tastes in different regions and for cooking mushrooms one has to follow his or her own style.

Khader and Padmavati (1991) observed that dried fruiting bodies were more nutritious than immature and mature stages of Oyster mushrooms. The people preferred the mushroom pickle, chutney powder and ketchups and provided methodology for the preparation of these.

Das (2003), based on oil and spices used, demarcated mushroom dishes into South and North Indian dishes. He reported standardization of about 60 different recipes at Department of Plant Pathology, Vellayani. He provided methodology for the preparation of vegetarian and non-vegetarian dishes of mushrooms mainly from the Kerala state.

Lot of work on this aspect has been done in the Department of Biosciences, Himachal Pradesh University, Shimla. Lakhanpal and his co-workers (unpublished) from time to time collected data on the ways and methodology of consumption of wild edible mushrooms. Shad (1985) gathered information on local recipes of morels, which were popular among the native and tribal people of Himachal Pradesh. Similarly, Kaisth (1985) provided information on the collection and traditional consumption of two species of *Lactarius*, viz. *L. delicious*, and *L. sanguifluus* in different parts of H.P. He described twelve most prevalent traditional recipes among inhabitants of Himachal Pradesh. In a similar type of study, Sharma (1987) reported consumption of two different species of *Sparassis*, viz. *S. radicata* and *S. crispa* and provided data on the popular recipes for this mushroom as used by the native people of H.P.

Sharma (1993) collected data on consumption and described the recipes for the preparation of five traditionally consumed mushrooms, viz. dry vegetable, mixed vegetable, 'rayta', 'pulao' and soups. Later Sharma (2001) collected the data from Hamirpur for the preparation of various wild mushrooms being consumed there. She described the recipes for dry vegetable, mixed vegetable, koftas, 'pakore' and 'poori' of *Termitomyces microcarpus*, *Leucoagaricus* sp., *Macrolepiota procera*, and *Collybia* sp., etc. Likewise, Rathore (2001) gathered data on some traditional and non-traditional dishes of *Morchella* species prevalent among the native people of Chamba district of H.P., especially the tribal belt. Similar studies were conducted by Rana (2002) on recipes of morels in the tribal district of Kinnaur. She listed some traditional and modern recipes of *Morchella*. In another study, Chauhan (2003) listed edibility of local common wild edible mushrooms of Kinnaur and described manners and methods of consumption such as dry vegetable, mixed vegetables and soups.

In this compilation similar data on a much larger scale has been gathered on the recipes of morels and the same is being presented in the following pages.

OBSERVATIONS

Recipes of Morels

While doing PRA (Participatory Rural Appraisal) exercise to record the ethnomycological data related to morels, the native people were interviewed to gather information on the preparation and consumption of morels. Following recipes were observed to be popular among the native people of the tribal belt.

I. Traditional Recipes of *Morchella* Species

(A) Rice preparation

1. Guchhi Ka Bhat/Rangmutz ka Bhat (Pulao of *Morchella*)

Ingredients:

Dried morel fruiting bodies	-	100 g
Basmati rice	-	2 cups
Onions	-	2 (small size)
Tomatoes	-	2 (medium size)
Garlic paste	-	3 flakes
Ginger paste	-	1 tbsp
Cloves	-	2-3 No.
Cinnamon	-	2 No.

Coriander	-	1 tsp
Cumin seeds	-	1 tsp
Turmeric	-	½ tsp
Refined oil	-	2 tbsp

Salt as per taste

The fruiting bodies of morels and rice are washed and soaked in water for half an hour. The refined oil is heated in a pan, cumin seeds and coriander are added to it, followed by onions and tomatoes and all ingredients are fried till they turn reddish brown. When the paste is ready, add to it turmeric powder, ginger–garlic paste, cloves and cinnamon along with chopped fruiting bodies of morels. The entire mass is stirred and allowed to simmer for 3-5 minutes. Now pre-soaked rice is added along with two cups of water, steamed for 5-10 minutes and then served with curd or 'desi ghee'.

2. Guchhi Biryani (*Morchella* Biryani)

Ingredients:

Dried morel fruiting bodies	-	100 g
Basmati rice	-	200 g
Onions	-	2 (small size)
Cloves	-	2-3 No.
Cinnamon	-	2 No.
Cardamom	-	2 No.
Cumin seeds	-	1 tsp
Bay leaves	-	2-3 No.
Refined oil	-	2 tbsp
Green coriander	-	2 tbsp

Salt and red chillies as per taste

The ingredients like cardamom, cloves, and cinnamon are ground coarsely. The dried fruiting bodies are chopped, soaked and then boiled in water. These are then boiled with the grounded ingredients and the decoction is sieved. The rice is half cooked and mixed with the sieved material in a pan with hot oil, fried onions and bay leaves. All the ingredients are steamed for 4-5 minutes and garnished with green coriander leaves while serving (Plate–II l).

3. Guchhi Pulao (Morel mixed veg Pulao)

Ingredients:

Dried morel fruiting bodies	-	100 g
Basmati rice	-	2 cups
Shelled peas	-	50 g
Onions	-	2 (medium size)
Capsicum	-	2 (medium size)
Cumin seeds	-	1 tbsp
Refined oil	-	2 tbsp
Green chillies	-	2 No.
Green coriander	-	2 tbsp

Salt and pepper as per taste

The fruiting bodies of morel and the basmati rice are washed and soaked in water for half an hour. After complete rehydration each fruiting body is sliced into 2-3 pieces. Separately in the pan are added cumin seeds and sliced onions and fried in oil till they turn deep brown. To this paste are added chopped morels, shelled peas, green chillies, salt and pepper. The soaked rice in two cups of water is added. The entire mixture is steamed for 5-10 minutes, garnished with green coriander leaves and served with curd or any pickle.

4. *Morchella* fried rice

Ingredients:

Dried morel fruiting bodies	-	200 g
Onions	-	1 (large size)
Basmati rice	-	3 cups
Cumin seeds	-	1 tbsp
Turmeric powder	-	A pinch
Green coriander	-	2 tbsp
Salted butter	-	5 tbsp

Salt and pepper as per taste

The fruiting bodies are soaked in luke warm water for half an hour for rehydration. The cumin seeds, sliced onions, turmeric and coriander powder are heated in a pan containing butter. Then the sliced fruiting bodies are mixed and cooked for 3-5 minutes. To this is added the pre-cooked rice. Salt is added as per taste. The entire mixture with ingredients is allowed to simmer for 3-5 minutes. While serving, it can be seasoned with pepper powder (Plate– IIk).

(B) Vegetable preparation

1. Sukhi subzi of Rangnu/Guchhi (Pure dry vegetable of *Morchella*)

Ingredients:

Morel dried fruiting bodies	-	200 g
Onions	-	2 (large size)
Tomatoes	-	2 (medium size)
Garlic paste	-	1 tsp
Ginger paste	-	1 tbsp
Green chillies	-	2 No.
Cloves	-	3-4 No.
Cinnamon	-	2 No.
Coriander powder	-	1 tsp
Cumin seeds	-	1 tsp
Turmeric	-	½ tsp
Refined oil	-	2 tbsp
Green coriander	-	1 tbsp

Salt and pepper as per taste

The fruiting bodies are washed and soaked in water for half an hour. The soaked fruiting bodies are sliced into two halves longitudinally. Separately in a pan, cumin seeds, chopped onion and tomatoes are fried in refined oil till they turn reddish-brown. When the paste gets ready, turmeric and coriander powder, cinnamon, cloves and ginger-garlic paste along with sliced fruiting bodies is added to the pan. All the ingredients are mixed properly and steamed for 5-10 minutes. The preparation is garnished with chopped green coriander leaves and served with 'chapati' or rice.

2. Mixed Vegetable of *Morchella* (Missi Kissi subzi of Guchhi)

Ingredients:

Dried morel fruiting bodies	-	200 g
Shelled peas	-	100 g
Cauliflower	-	50 g
Cabbage	-	50 g
Onions	-	2 (medium size)
Tomatoes	-	2 (medium size)
Garlic paste	-	1 tsp
Ginger paste	-	1 tbsp
Coriander powder	-	1 tsp
Cumin seeds	-	1 tsp
Turmeric powder	-	½ tsp
Refined oil	-	2 tbsp
Green coriander	-	1 tbsp

Salt and chillies as per taste

Fruiting bodies are soaked in salted water for half an hour and then each sliced into 2-3 pieces. Cumin seeds, sliced onions, tomatoes and ginger-garlic paste are fried in refined oil till the mixture turns reddish-brown. To this mixture, turmeric and coriander powder and chopped vegetables are added and all the ingredients are mixed properly; little water can be added, if necessary. The entire mixture is cooked for 8-10 minutes. The preparation can be garnished with green coriander leaves before serving (Plate–1a).

3. Curry of *Morchella* (Curry of Guchhi)

Ingredients:

Dried morel fruiting bodies	-	100 g
Onions	-	2 (large size)
Garlic	-	2–3 flakes
Ginger	-	5 g
Cumin seeds	-	1 tsp
Coriander powder	-	1 tsp
Turmeric	-	½ tsp
Mustard oil	-	2 tbsp
Green coriander	-	1 tbsp
Curd	-	2 cups
Curry leaves	-	A few

Salt and chillies as per taste

The dried fruiting bodies of morels are soaked in water for half an hour. Cumin seeds,

PLATE–I (a–f) (a) Mixed vegetable of *Morchella* (Missi-kissi subzi of guchhi) with ingredients (turmeric powder, tomatoes, potatoes, guchhi, green chillies, onion, salt, oil, mixed vegetables). **(b) *Morchella* flavoured dish** (Tardar subzi of guchhi) with ingredients (turmeric powder, tomatoes, potatoes, guchhi, green chillies, onion, salt, oil). **(c) Koftas of *Morchella*** with ingredients (turmeric powder, tomatoes, guchhi, green chillies, onion, salt, oil). **(d) Mixed-veg soup of *Morchella*** (ingredients – vegetables: cabbage, capsicum, tomatoes, onion, turmeric, garlic, guchhi, green chillies, salt, refined flour, butter, cheese, green coriander). **(e) Curry of *Morchella*** (curry of guchhi) (ingredients–turmeric powder, guchhi, green chillies, onion, salt, oil, curd). **(f) Peas *Morchella*** (guchhi matar) with ingredients (turmeric powder, tomatoes, potatoes, guchhi, green chillies, onion, salt, mustard oil, peas).

red chillies, sliced onions and garlic are fried in vegetable oil or butter till the mixture turns golden-brown. To this paste turmeric and coriander powder, chopped fruiting bodies, grated ginger are added and allowed to simmer. Then curd, curry leaves and little water is added to this paste. All the ingredients are mixed properly and stirred for 15-20 minutes. The preparation is garnished with green coriander leaves (Plate–Ie).

4. *Morchella* Shahi Paneer

Ingredients:

Morel fruiting bodies	-	200 g
Paneer	-	100 g
Onions	-	2 (large size)
Tomatoes	-	2 (medium size)
Garlic	-	2-3 flakes
Ginger	-	1 (small size)
Cumin seeds	-	1 tsp
Turmeric	-	½ tsp
Coriander powder	-	1 tsp
Refined oil	-	2 tbsp
Green coriander	-	1 tbsp
Salt as per taste		

The fruiting bodies of morel are washed and soaked in water for half an hour and then sliced. Cumin seeds, garlic, chopped onions are fried in oil till they turn golden-brown. After allowing the mixture to simmer for 3-4 minutes, cheese (paneer) slices and chopped pieces of mushrooms are added. All the ingredients are heated in little water. While serving, the recipe is garnished with green coriander leaves.

5. Guchhi Masala (Morel Masala)

Ingredients:

Morels shredded	-	250 g
Green chillies	-	2 (medium size)
Onions	-	2 (large size)
Ginger shredded	-	1 tsp
Garlic	-	3 flakes
Clove	-	2 No.
Cardamom	-	2 No.
Cinnamon	-	2 No.
Cumin seeds	-	1 tbsp
Turmeric powder	-	A pinch
Oil	-	For frying
Curry leaves	-	A few
Salt as per taste		

The fruiting bodies of morels are rehydrated in water, and then to the oil in a 'kadai' (saucer pan), cumin seeds are added. Side by side chopped onions, ginger, green chillies and curry leaves are marinated with the shredded mushrooms. To this marinated mixture, sprinkle water if necessary. The mixture is cooked for 10-12 minutes and served hot with 'chapati' or rice (Plate–IIg).

6. Taridar subzi (Gravy of vegetable) of Guchhi (*Morchella* Flavoured Dish)

Ingredients:

Morel fruiting bodies	-	250 g
Grinded tomatoes	-	100 g
Grinded onions	-	50g
Corn flour	-	1 tbsp
Cumin seeds	-	1 tsp
Coriander powder	-	1 tsp
Turmeric	-	1 tsp
Green coriander leaves	-	2 tsp
Refined oil	-	2 tbsp

Salt as per taste

The morels fruiting bodies are soaked in water and split into halves. Cumin seeds, chopped onion and tomatoes are fried in oil. The gravy is thickened by adding corn flour. The mixture is stirred continuously. To this mixture, turmeric, coriander powder and water are added and the mixture is allowed to simmer for 3-5 minutes. Salt and chillies are added as per taste. Fruiting bodies are added and fried and all the ingredients steamed for 15-20 minutes and the preparation is garnished with green coriander leaves (Plate–Ib).

7. 'Madra' of Guchhi (Madra of *Morchella*)

Ingredients:

Morel fruiting bodies	-	200 g
Curd	-	2 cups
Ginger paste	-	1 tbsp
Garlic paste	-	1 tsp
Turmeric powder	-	½ tbsp
Coriander powder	-	1 tsp
Cumin seeds	-	1 tbsp
Onion	-	1 (medium size)
Tomato gravy	-	1 cup
Refined oil	-	5 tbsp
Green coriander leaves	-	2 tbsp

Salt and pepper as per taste

Fruiting bodies of morels are washed and soaked for half an hour. The rehydrated fruiting bodies are deep-fried in the refined oil and kept aside. Cumin seeds, chopped onions and tomatoes are fried separately. The curd is added and the mixture is continuously stirred for 20-30 minutes till the oil separates from curd. Then garlic and ginger paste is made and added along with the fried fruiting bodies and the mixture is steamed for 10-20 minutes. It

PLATE–II (g–n) **(g) Morel masala** (Guchhi masala) (ingredients – tomatoes, onion, potatoes, turmeric powder, garlic, guchhi, green chillies, salt, green coriander). **(h) *Morchella* omelette** (ingredients – eggs, morels, green chillies, butter, refined oil, salt). **(i) *Morchella* snacks** (ingredients – bread, refined oil, chillies, shredded morels, onions, gram flour, salt). **(j) *Morchella* toast** (ingredients – brown bread, butter, potato stuffing (tikki), cabbage, tomatoes, cucumber slices, shredded morels). **(k) *Morchella* fried rice** (Guchhi ka bhat) (ingredients – tomatoes, onion, turmeric powder, garlic, guchhi, green chillies, salt, oil, basmati rice, green coriander). **(l) Biryani of morels** (ingredients – tomatoes, onion, vegetables, turmeric powder, garlic, guchhi, green chillies, bay leaves, cardamom, cinnamon, oil, salt, basmati rice, green coriander). **(m) Stuffed vegetarian morels** with ingredients (vegetables – cabbage, capsicum, tomatoes, onion, turmeric, garlic, green chillies, onion, salt, oil). **(n) Non-veg soup of morels** (ingredients – chicken boneless pieces, refined flour, cheese, butter, salt, black pepper, garlic and ginger)

can be garnished with green coriander leaves while serving.

8. Matar Guchhi (Peas *Morchella*)

Ingredients:

Dried morel fruiting bodies	-	200 g
Shelled Peas	-	500 g
Onion	-	2 (large size)
Tomatoes	-	2 (medium size)
Garlic paste	-	1 tsp
Ginger paste	-	1 tbsp
Green chillies	-	2 No.
Coriander powder	-	1 tsp
Cumin seeds	-	1 tsp
Turmeric powder	-	½ tsp
Refined oil	-	2 tbsp

Salt and pepper as per taste

The dried fruiting bodies are washed and soaked for half an hour and then sliced into small pieces. Cumin seeds, chopped onions, tomatoes, turmeric-coriander powder, and ginger-garlic are fried in refined oil till they turn reddish-brown. The sliced pieces of morels and shelled peas are added to this mixture and steamed for 5-10 minutes. The preparation can be served with 'chapati' or rice (Plate–If).

9. *Morchella* 'Omelette'

Ingredients:

Morels finely chopped	-	100 g
Eggs	-	5 No.
Grated cheese	-	5 g
Butter	-	25 g
Green coriander	-	1 tbsp

Salt and pepper as per taste

The fruiting bodies are washed and soaked in luke warm water for half an hour, sliced into small pieces and then fried in butter. Simultaneously, the eggs are beaten in a bowl. Salt, pepper and green coriander leaves are added along with grated cheese. All the ingredients are mixed properly and then deep fried in butter (Plate–IIh).

10. 'Kofta' of Guchhi (Morel Koftas)

Ingredients:

Dried morel fruiting bodies	-	200 g
Red chillies	-	½ tsp
Grinded tomatoes	-	2 (small size)
Grinded onions	-	2 (medium size)
Turmeric powder	-	½ tsp
Coriander powder	-	1 tsp
Grinded anardana	-	2 tsp

Gram flour	-	4 tbsp
Refined oil	-	4 tbsp
Salt as per taste		

The fruiting bodies are soaked in salted water and the mixture is sieved. To the fruiting bodies, coriander powder, grinded *anardana* (*Pomegranate* seeds), red chillies, salt (as per taste) along with gram flour are added and the contents are mashed and made into small round balls, which are deep-fried in refined oil. Side by side, onions, tomatoes, garlic and ginger are made into a paste. These 'stuffed balls', called 'koftas', are added to it. The gravy is boiled for 10-15 minutes and garnished with green coriander (Plate–Ic).

(C) Soups

1. Soup of *Morchella* (Khadha of Guchhi)

Ingredients:

Dried morel fruiting bodies	-	200 g
Ginger	-	10 g
Water	-	500 ml
Desi makkhan (butter)	-	2 tbsp
Salt to taste		

The fruiting bodies are boiled in salted water with salt for half an hour. The decoction is flavoured with chopped ginger. The mixture is sieved and served hot by adding butter/soy sauce or chilly sauce.

II. Modern Recipes of *Morchella* Species

For gathering information on the modern recipes of *Morchella* species, some posh families, restaurants and hotels were visited and information gathered with regard to the recipes. The morels recipes were reported to be the costliest dishes in hotels and restaurants.

(A) Soups

1. Soup of *Morchella*

Ingredients:

Dried morel fruiting bodies	-	50 g
Onions	-	1 (small size)
Ginger	-	5 g
Garlic	-	2 flakes
Refined flour	-	1 tbsp
Butter	-	2 tbsp
Grated cheese	-	1 tbsp
Green coriander	-	1 tbsp
Salt and pepper as per taste		

The fruiting bodies of morel are rehydrated in water and then finely sliced. To the pieces of morels, chopped onions, grated ginger, garlic are added and boiled for 20-30 minutes.

Separately in a pan, refined flour is fried in melted butter. The decoction is sieved and boiled for 5-10 minutes till the volume of the contents is reduced to one half. Pepper and salt are added as per taste and garnishing is done with green coriander leaves.

2. Mixed Veg Soup of *Morchella*

Ingredients:

Dried morel fruiting bodies	-	100 g
Capsicum	-	2 No.
Radish	-	5 g
Onion	-	1(small size)
Ginger	-	3 g
Garlic	-	2 flakes
Refined flour	-	1 tbsp
Butter	-	2 tbsp
Grated cheese	-	1 tbsp
Green coriander	-	1 tbsp

Salt and pepper as per taste

The washed, rehydrated and chopped fruiting bodies are mixed with ingredients like *Capsicum*, radish and onion. These ingredients are mixed with refined flour and fried in butter. Water is added and the entire mixture is stirred continuously for 30-35 minutes. Salt and pepper are added to taste and the soup is garnished with grated cheese and green coriander leaves (Plate–Id).

3. Mixed Non-Veg Soup of *Morchella*

Ingredients:

Dried morel fruiting bodies	-	100 g
Keema	-	100 g
Milk	-	3 cups
Butter	-	30 g
Refined flour	-	1 tbsp
Grated cheese	-	1 tbsp

Salt and pepper as per taste

The fruiting bodies of rehydrated and chopped morels and meat stock are boiled for 20-30 minutes. These are mashed and sieved in a muslin cloth. To the butter in a pan, refined flour is fried in butter and then a mixture of milk, mushroom and meat stock is added. Salt is added as per taste and the ingredients are boiled for 20-30 minutes. For flavour freshly grinded pepper is added (Plate–IIn).

4. ***Morchella*** **Mycelium Flavoured Soup**

Ingredients:

Morchella mycelium	-	50 g
Onions	-	1 (small size)
Ginger	-	5 g
Garlic	-	2 flakes
Refined flour	-	1 tbsp

Butter - 2 tbsp
Grated cheese - 1 tbsp
Green coriander - 1 tbsp
Salt and pepper as per taste

The fresh mycelium of morels is mixed with chopped onion, grated ginger and garlic Separately refined flour is fried in butter. All the ingredients are then mixed properly. The mixture is boiled for 5-10 minutes till the volume remains half. Pepper and salt are added as per taste and garnished with chopped green coriander leaves.

(B) Stuffings

1. Stuffed Non-vegetarian Morels

Ingredients:

Large sized morel fruiting bodies - 15 No.
Keema - 100 g
Tomatoes - 1 (large size)
Onions - 2 (medium size)
Paste garlic - 1 tsp
Grated ginger - 1 tbsp
Turmeric powder - ½ tbsp
Coriander powder - 1 tsp
Cumin seeds - 1 tbsp
Red chillies, salt, asafoetida as per taste

The dried fruiting bodies are washed and soaked in water and split longitudinally. Cumin seeds are fried in oil and added to a paste of onions, tomatoes, ginger and garlic paste and fried till they turn golden-brown. Now turmeric, coriander powder and keema are added along with salt, chillies and asafoetida. The ingredients are boiled for 5-7 minutes. This mixture is then stuffed into fruiting bodies of morels. The stuffed fruiting bodies are wrapped in aluminium foil and baked for 10-15 minutes.

2. Stuffed Vegetarian Morels

Ingredients:

Large sized fruiting bodies - 15 No.
Cabbage - 10 g
Capsicum - 2 No.
Grated carrot - 2 No.
Chopped tomatoes - 1 (large size)
Chopped onions - 2 (medium size)
Paste garlic - 1 tsp
Grated ginger - 1 tbsp
Turmeric powder - ½ tbsp

Coriander powder - 1 tsp
Cumin seeds - 1 tbsp
Red chillies and salt as per taste

The dried fruiting bodies are soaked in water and split longitudinally. Cumin seeds are fried in oil along with paste of onions, tomatoes, ginger and garlic and fried till they turn golden brown. After that turmeric, coriander powder and finely chopped vegetables are added to this mixture. Salt, chillies and asafoetida are added as per taste. The entire mixture is simmered for 5-7 minutes. This mixture is stuffed into fruiting bodies of morels. The stuffed fruiting bodies are wrapped in aluminium foil and baked for 10-15 minutes (Plate–IIm).

Snacks

1. *Morchella* Snacks

Ingredients:

Morels shredded - 100 g
Green chillies - 2 No.
Onion - 1 (large size)
Grated ginger - ½ tsp
Gram flour - 5 tbsp
Turmeric powder - A pinch
Refined oil - 50 ml
Salt and pepper as per taste

The gram flour is kneaded uniformly with hot water and the mixture is made into dough. For stuffing, onion, chillies, ginger and shredded morels are sautéed in a oil. Salt and pepper are added as per taste, and the mixture is cooked properly. The dough is spread and filled with the mixture. The edges are folded and fried in oil. These snacks are served with 'chutney' or sauce (Plate–IIj).

2. *Morchella* Snacks with Egg

Ingredients:

Morel fruiting bodies - 100 g
Eggs - 4 No.
Green chillies - 2 No.
Grated ginger - ½ tsp
Gram flour - 5 tbsp
Refined oil - 50 ml
Salt and pepper as per taste

The morel fruiting bodies are rehydrated by soaking in water and then sliced into two longitudinal halves. The eggs are beaten in a bowl and to it are added shredded chillies, grated ginger and the salt and pepper as per taste. The pieces of the fruiting bodies are again beaten with the egg mixture. The oil is heated and one by one, the pieces are rolled evenly in the flour, so that the flour gets attached to the egg mixture evenly. These are then fried till they turn brown. These can be served with 'chutney' or sauce (Plate–IIi).

3. Morels in White Wine

Ingredients:

Morel fruiting bodies	-	10 (medium size)
Garlic paste	-	2 tbsp
Refined oil	-	2 tbsp
Butter	-	100 g
Parsley	-	2 tbsp

Salt and pepper as per taste

The dried fruiting bodies of morel are rehydrated in water and then finely sliced into 2-3 pieces each. Garlic paste is sauté with rehydrated morels in the combination of oil and butter and fried till the morels turn light brown. Now the pan is filled with white wine along with freshly chopped parsley. The mixture is simmered for few minutes and served after adding freshly ground pepper.

Morels Pickle

Ingredients:

Fresh morels	-	20 g
Red chillies	-	2 g
Mustard seeds	-	3 g
Garlic (chopped)	-	5 g
Red chilli powder	-	5 g
Turmeric powder	-	3 g
Mustard seeds powder	-	10 g
Salt	-	20 g
Mustard oil	-	250 ml

The fruiting bodies are sliced and fried in mustard oil and are seasoned with red chillies and mustard seeds along with mustard seed powder, salt and fried morels. All the ingredients are mixed properly and stored in a clean bottle. The mixture is kept for few days and then the pickle is ready for consumption.

Annexure

ANNEXURE–I

Methodology for Collection, Preservation and Identification of Morels

a) *Collection*

The morels predominately fruit in spring season[1]. The fruiting season extends from February–March to May–June with its maxima in March–April. Regular field forays were carried out to different parts of Himachal Pradesh. The field trips varied from whole day treks to weekly trips to the nearby areas, whereas the distant sites were surveyed by temporarily camping there and then different localities in the vicinity were explored. The collection was done as given below:

(i) The sporophores were picked up individually or collectively if they occurred gregariously, wrapped up in the wax paper, placed in the collection bags, and carefully brought to the base camp. Efforts were made to collect fruit bodies at different developmental stages. Habit, habitat association and relevant ecological data like altitude, forest type, temperature and humidity were all recorded in the field.

(ii) In the laboratory or base camp, the collections were spread out and each sample was allotted a collection number.

b) *Preservation and drying of the material*

The specimens were preserved dry as well as wet.

(i) Dry preservation: The collections were sun dried by spreading them out on filter papers. The dried specimens were treated with (1:1000 W/V) mercuric chloride to prevent insect damage and dried again. The collections were kept in paper bags, which were wrapped in polythene bags and sealed properly. These bags were stored in damp-proof conditions in the herbarium almirahs.

(ii) Wet preservation: Wet preservation was done in alcohol–formalin[2] in glass jars following Ainsworth (1971). For preservation, specimen jars with bakelite screw caps were used. The specimens were kept for 24 hours in this solution and permanently placed in 50 per cent alcohol.

c) *Macroscopic observations*

The various macroscopic characters, which helped in the identification of morels, were recorded in the fresh specimens in the field. For this purpose scientific keys were designed highlighting the following diagnostic macroscopic features. These characteristics were carefully recorded as in the key given below:

---

[1]Observations and variations in the fruiting period have been recorded in the Chapter 5 on Ecological aspects of morels at p. 83–89.

[2]Formula for fixative (Ainsworth, 2001)

Formalin 40%	-	25.0 ml
Alcohol 95%	-	150.0 ml
Water	-	1000 ml

Pileus: Shape, size, colour of pits, arrangement and colour of ribs.

Stipe: Shape, size, colour, smooth or lacunose, swollen at the base or not.

The colour was recorded following Methuen Handbook of colour (Kornerup and Wanscher, 1978).

d) Microscopic studies

For making microscopic observations both dried as well as wet specimens were utilized. The dried sporophores were revived by soaking in 2% aqueous solution of potassium hydroxide, whereas the preserved specimens were used directly after washing with water. Microscopic details of the specimens were worked out in laboratory with the help of research microscope (Getner Binocular Research Microscope GTRbA). Measurements of asci, ascospores, paraphyses and other parts have been represented by the scale. All the microscopic structures were drawn with the help of Camera Lucida. The magnifications of the system with various combinations of eye pieces and objective lenses were x 2350 (with 10 x 100), x 1040 (10 x 45) and x 235 at 10 x 10. The following microscopic features were studied in detail.

- **Spores:** The ascospores were studied from the tissue of pits. The spores were mounted in water and observed under microscope (Dennis, 1968). Spore size and shape are given as seen in lateral view. The apothecial tissue was mounted in Melzer's reagent[3] (Melzer, 1934) and examined under oil immersion for their size and ornamentations.
- **Hymenium:** The asci and paraphyses were studied by tearing out the apothecial tissues as by cutting free hand sections through apothecia from the preserved dried material. The sections were stained in one per cent aqueous Phloxine[4] solution Sections were also mounted in Lactophenol, stained in cotton blue[5] in Lactic acid and then mounted in Lactophenol[6].
- **Hair:** Hairs were studied by taking tissues from the stipe and mounting it in the same way as asci and paraphyses.

---

[3]**Melzer's Reagent:**	Potassium Iodide	-	1.5 gm
	Iodine	-	0.5 gm
	Chloral hydrate	-	22.0 gm
	Distilled water	-	22.0 ml
[4]**1% aqueous Phloxine:**	Phloxine	-	1.0 gm
	Water	-	99.0 ml

One or two minutes and mounted in slightly acidified 50 per cent aqueous Glycerine ++ for routine examination.

[5]**50% aqueous Glycerine**	Glycerine	-	50.0 ml
	Water	-	50.0 ml

[6]Saturated solution of cotton blue in spirit – 10 ml

	Pure glycerine	-	10 ml
	Water	-	80 ml
[7]**Lactophenol**	Lactic acid	-	10 ml
	Phenol	-	10 ml
	Glycerine	-	10 ml

ANNEXURE–II

Methodology for Genotypic Polymorphism and Diversity

Protocol for isolation of genomic DNA from mushroom (Ausubel et al., *1995)*

Two g of *Morchella* sporophores were grinded in liquid Nitrogen to fine powder in a pre-chilled pestle and mortar. The powdered mushroom material was transferred to a 50 ml centrifuge tube containing 14 ml of prewarmed (65°C) extraction buffer and mixed by gentle inversion. This was incubated at 65°C for one hour. Fourteen ml chloroform: isoamyl alcohol (24:1, v/v) were added and mixed by inversion. The spinning was done at 10,620 x g for 20 minutes at room temperature.

The aqueous phase was pipetted out into another tube without disturbing the interphase. 2/3rd volume of isopropanol was added and mixed by gentle inversion and spined down the DNA pellet. The pellet was washed in 70% ethanol and spined at 10,620 x g for 10 minutes at 4°C. It was then dried and dissolved in 100 μl of sterile H_2O. Ten μl of RNAase (10 mg/ml) was added to the final concentration of 10 μg/ml, mixed gently and incubated for 30 min/1 hour at 37°C. An equal volume of phenol: chloroform: isoamyalalcohol (25:24:1) was added, mixed thoroughly and spined at 10,620 x g for 10 minutes. The upper aqueous phase was carefully taken using a 200 μl pipette without disturbing the interphase. 1/10 volume of 3M sodium acetate, pH 5.2 and 2 volumes of absolute ethanol (100%) was added and mixed by gentle inversion, and spined at 10,620 x g for 10 minutes to pellet the DNA. The pellet was washed with 70% ethanol and dissolved in sterile water.

Fungal isolates

Seven different species of *Morchella* were analyzed and used in this study. The isolates and their geographical sites are listed in Table 4.1. The investigated taxa were *Morchella angusticeps* Peck, *M. conica* Fries, *M. crassipes* Vent., *M. deliciosa* Fries, *M. esculenta* L., *M. tibelica* and *M. simlensis* sp. nov. (Lakhanpal and Shad, 1999).

DNA isolation and purification of Morchella *species*

Total genomic DNA from the mushroom fruiting bodies was extracted by Cetyl-Trimethyl Ammonium Bromide (CTAB) procedure (Doyle and Doyle, 1987). RNA contaminants in all the samples were digested with 100 mg ml^{-1} RNAase for 30 minutes at 37°C, extracted once with phenol: chloroform: isoamyl alcohol (25:24:1), precipitated by adding 0.1 volume ammonium acetate and 2 volume of chilled ethanol, washed with 70% ethanol and resuspended in 200 ml of TE buffer (10 mM Tris.Cl pH 8.0 and 1mM EDTA pH 8.0). Quantity of DNA in each sample was estimated by measuring absorbance at 260 nm in a spectrophotometer (SmartSpec 3000 of Bio-Rad). The quality of DNA preparation was checked both by taking A_{260}/A_{280} ratio as well as by gel electrophoresis. A_{260}/A_{280} value between 1.8–2.0 was taken as good quality DNA preparation. If the ratio was below 1.7, the preparation was purified once again by phenol: chloroform extraction procedure. Finally, all the DNA samples were diluted in sterile MilliQ water so that they contained 25 ng DNA / ml (for detail protocol see Annexure –II).

Random primers used for RAPD analysis

RAPD amplifications were performed as per Williams *et al.* (1990). The random primers used in this study were 10 nucleotide long. Sequences of each primer were selected randomly keeping in view that the GC content is 60-70% and that the primer ends are not self-complementary. These were purchased from M/S Operon Technology, Inc., 1000 Atlantic Ave, Alameda, CA 94501, USA. A total of 28 primers were used in this study (Chapter 4).

DNA Amplification

Polymerase Chain Reaction (PCR) was carried out in a reaction volume of 25 µl containing 10mM Tris-HCl (pH 8.3), 50 mM KCl, 0.1% Triton X 100, 1.5 mM $MgCl_2$, 200ìM each dNTPs, 25 pmol primer, 100 ng genomic DNA and 1 unit of Taq DNA polymerase (Ampli Taq, Perkin Elmer). Amplification was carried for 45 cycles and each cycle consisted of denaturation at 94°sC for 1 minute, annealing at 35.5°C for 1minute and 2 minutes extension at 72°C. Cycles were preceded by denaturation for 5 minutes at 94°C, followed by a final extension at 72°C for 10 minutes.

Electrophoresis

After amplification, the products were separated by electrophoresis in 1.6% Agarose gel (in Tris borate EDTA buffer) with 0.5 μgml^{-1} ethidium bromide. The gel was run at 5V/cm until the loading dye reached the gel front. Gel image in which bands represented the DNA strands was captured in UV light and stored in a computer using gel Fluor-S^{TM} Multi-imager (Bio-Rad). DNA fragments were detected and aligned by Diversity Database Software (Bio-Rad). The amplification reactions for each primer were repeated twice to ensure reproducibility, and only those bands that occurred consistently were considered for the analysis.

Statistical analysis of data

Data recorded for genotypic analysis in *Morchella* spp. were statistically analysed. All gels were scanned and stored as TIFF (Tagged–Image File Format) files. RAPD products were scored in binary fashion as (0) indicating absence and (1) indicating presence of a band for each primer. The analysis of genetic similarity among seven *Morchella* species was done using NTSYS - pc (version 2.0) software (Rohlf, 1998), using dice coefficient (Nei and Li, 1979). The output data clusters were graphically represented by phylogenetic tree (dendrogram).

ANNEXURE–III

Methodology for Data on Ecological Parameters

Four sites where morels appeared regularly were marked and selected for detailed study – (i) Jalori pass, (ii) Nohnu pass, (iii) Chazpur, and (iv) Narkanda. For collecting morels from different forest types and altitudinal and vegetational zones, the selected forests were surveyed along a gradient (1600–2800 m) starting from lower altitude and travelling up to higher altitude. Along such gradient, the floristic composition changed gradually after almost every 500 m making it possible to collect morels from different forest types and altitudinal and vegetational zones. For recording the fruiting period of morel species, the sites were visited regularly from mid-February to June. The colour of the fructification and other key characters of the species were recorded in the field as given in Chapter 3 (p - 41). Data on temperature, moisture, pH of the substrate and relative humidity were also regularly recorded. The temperature was recorded by thermometers placed at (i) ground level and (ii) at 0.12 cm below ground level. The former represented the surface temperature, (i.e. of emerging sporophores) and the latter the temperature of the mycelial region where sporophores were being presumably formed.

Soil Analysis

Soil samples in triplicate, from specific morel bearing and non-morel bearing soils, were collected from different forest types from each species of *Morchella*.

The soil samples were dried at 60°C in oven and then ground to powdered form.

(i) Soil pH was recorded with digital pH meter by dissolving 5 gm of soil in 25 ml of distilled water.

(ii) Soil colour was noted visually and compared with Methuen Hand Book of Colour (Kornerup and Wanscher, 1978).

(iii) Nitrogen was estimated by Kjeldahl digestion method (Trivedi and Goel, 1984) as follows:

Ten gm of the soil was taken in 300 ml Kjeldhal flasks. To this were added 25 ml of distilled water, 20 gm of catalyst mixture (K_2SO_4 – 400 gm + $CuSO_4$ - 20 gm + HgO – 3 gm + Selenium powder – 1 gm), and 35 ml Conc. H_2SO_4. Flasks were left for digestion till the solution was clear white. To this, 100 ml of distilled water was added, and supernatant was transferred in 100 ml distillation flask. To the supernatant was added 100 ml of 40 per cent NaOH. After distillation, ammonia gas was trapped in 500 ml conical flask containing 25 ml of boric acid and mixed indicator. Contents were titrated with 0.1N HCl. Readings were compared with standard chart for equivalent percentage of nitrogen. The percentage of nitrogen was calculated as follows:

$$\text{Nitrogen \%} = \frac{A \times 0.14 \times N \times 100}{W}$$

where, A = Volume of HCl used

N = Normality of HCl

W = Weight of the sample

iv) Estimation of Phosphorus was done following Olsen *et al.* (1954), in which 1.0 gm of soil was taken in 100 ml conical flasks. To this was added a pinch of Darco G 60 and 20.0 ml of 0.5 N sodium bicarbonate. Five ml of the filtrate was taken in 25 ml measuring flask. To this were added 0.5 ml of ammonium molybdate, 10 ml of distilled water and 1.0 ml of dilute stannous chloride solution. Distilled water was added to make the total volume upto mark. After half an hour, readings were taken with Spectronic - 20 at 660 nm.

For Calcium, Magnesium and Potassium estimation, 10 gm of fine textured soil was washed with 80 per cent alcohol 2-3 times and then air dried. To this was added 20 ml of ammonium acetate solution and stirred thoroughly. The suspension was kept as such for some time and filtered through Whatman–50 filter paper. The supernatant was treated as stock solution to determine Ca, Mg, Na, and K concentration. Na and K were determined by directly feeding the stock solution into the Corning 400–Flame photometer. The standard curves of Na and K were made by preparing standard solution of KCl and NaCl following Dean (1960). For Ca, 1 ml of stock solution was taken and a pinch of mercuric oxide indicator was added to it. The red coloured solution was obtained, which was titrated against EDTA (Ethylene diamine Tetraacetic acid). The change of pink colour to purple was treated as end point.

The concentration of Ca in sample was treated as end point. The concentration of Ca in sample was then calculated from the volume of EDTA used.

Mg and Ca were combinedly determined by taking 1 ml of stock solution, to which were added 1 ml of NH_4Cl - NH_4OH buffer and a drop of EBT (Erichrome Black dye) indicator. The red colour solution formed was titrated against 0.01 EDTA till the blue colour was obtained. From the volume EDTA used for both Ca and Mg, the volume for Ca was subtracted to determine Mg concentration.

ANNEXURE–IV A

Methodology for Experiments on Physiological Aspects

Isolation and establishment of pure culture

The pure cultures were raised from the pileus and stipular tissues of healthy, sun dried, fresh specimens. Young/dried specimens were first washed with distilled water and then the tissue of stipe or pileus regions were cut with the help of sterilized sharp blade. The tissues were surface sterilized by 0.1% mercuric chloride solution. Now the tissue was placed on filter paper to remove the excess moisture. The small bits of *Morchella* tissues were then transferred onto Potato Dextrose Agar (PDA) medium slants or Petri dishes aseptically with the help of sterilized inoculating needles. These were incubated at 22 ± 1° C for at least 6-10 days. The cultures were obtained after 6-10 days on the media. The actively growing mycelial colonies were subcultured to obtain pure cultures.

Media

Ten solid and four different liquid media were evaluated to find the most suitable medium for the optimum mycelial growth as given by Tuite (1969). The compositions of various media are also given on page no. (xiii).

Glasswares used

In all the experimental studies, the glasswares used were of Borosil. In the beginning of every experiment, the glasswares were first cleaned with washing reagent and then washed with tap water.

Sterilization

All glassware articles, e.g. Petri plates, test tubes, etc. were sterilized in oven at 175°C for two hours. The different solid and liquid media used in the present study were sterilized at 121.6°C (15 p.s.i.) for 20 minutes in autoclave. Ethyl alcohol was used for cleaning inoculating needle, cork borer and forceps. These were first hand dipped in ethyl alcohol and then flame sterilized. The inoculation was done in ultraviolet sterilized chamber—Laminar flow.

Inoculation

(i) Solid media*

Inoculum used in various experiments was obtained from the periphery of actively growing

---

*Composition of Solid media

1. **Malt yeast agar extract**
   - Agar-Agar 20 g
   - Malt extract 10 g
   - Yeast extract 1 g
   - Distilled water 1000 ml
2. **Malt Agar**
   - Agar-Agar 20 gm
   - Malt extract 25 gm
   - Distilled water 1000 ml
3. **Potato Dextrose Agar**
   - Agar-Agar 20 gm
   - Potato peeled 250 gm
   - Dextrose 20 gm
   - Distilled water 1000 ml
4. **Yeastal PDA**
   - Agar-Agar 20 gm
   - Potato peeled 250 gm
   - Dextrose 20 gm
   - Dextrose 20 gm
5. **Pea extract**
   - Crushed pea seeds 40 gm
   - Agar-Agar 20 gm
   - Distilled water 1000 ml
6. **Czapek's Dox Agar**
   - $NaNO_3$ 2 gm
   - K_2HPO_4 0.1 gm
   - $MgSO_4 . 7H_2O$ 0.5 gm
   - Sucrose 30 gm
   - Agar-Agar 15 gm
   - Distilled water 1000 ml

mycelial colonies on PDA medium. Five mm diameter discs of mycelia were taken out with a pre-sterilized cork borer under aseptic conditions to be used as inoculum in solid media for cultural studies. While cutting discs, care was taken to cut very thin portions of mycelial discs and to avoid agar blocks, as these can act as contaminating source.

(ii) Liquid media**

In the experiments on the liquid media, a small bit of mycelium was transferred to the flask containing liquid medium and incubated for 7-10 days at 20 ± 1°C . After 10 days, a mat of fungal mycelium appeared, which was homogenized in the medium by sterilized rod. Then equal amount of this homogenized mycelium was added to flasks having different liquid media.

Incubation

The Petri dishes and flasks having different solid and liquid media were incubated for 10-12 days at 20 ± 1°C in an incubator. The growth was periodically observed and mycelium was harvested at periodic intervals.

Recording of vegetative growth

Vegetative growth of mycelium in the solid media was measured by taking the diameter of colony in two directions at right angles. In the liquid media, the mycelium was filtered through oven dried weighed Whatman filter paper discs of 7.5 cm diameter. After filtration the mycelium along with filter paper was again oven dried at 70°C for 12 hours, and finally weighed to give dry weight of the mycelium. Three replicates of each mycelium were used and average values were taken for comparison of growth in different media.

---

7. Coriander extract

Coriander seeds	10 gm
Agar-Agar	20 gm
Distilled water	1000 ml

8. Maize grain extract

Crushed maize	40 gm
Agar-Agar	20 gm
Distilled water	1000 ml

9. Wheat grain extract

Glucose	1.0 gm
Crushed wheat grain	40 gm
Agar-Agar	20 gm
Distilled water	1000 ml

10. Glucose Yeast Agar

Glucose	1.0%
Yeast extract	0.3%
K_2HPO_4	0.2%
$MgSO_4 . 7H_2O$	0.2%
Agar-Agar	2.0%
Distilled water	1000 ml

**Composition of Liquid media

1. Glucose Asparagine

Glucose	30 gm
Asparagine	1 gm
K_2HPO_4	1.5 gm
Distilled water	1000 ml

2. Czapek's Dox solution

$NaNO_3$	2 gm
K_2HPO_4	0.1 gm
$MgSO_4 . 7H_2O$	0.5 gm
$KCl . 7H_2O$	0.01 gm
$FeSO_4 . 7H_2O$	0.5 gm
Sucrose	30 gm
Agar-Agar	15 gm
Distilled water	1000 ml

3. Dimmick's Solution

Glucose	10 gm
K_2HPO_4	0.5 gm
$MgSO_4 . 7H_2O$	0.2 gm
NaCl	0.5 gm

4. Asthana and Hawker'medium

Glucose	5.0 gm
Potassium nitrate	3.5 gm
$MgSO_4$	0.75 gm
KH_2PO_4	1.75 gm
Distilled water	1000 ml

The media with best vegetative growth were used in further studies, i.e. for studying the effect of temperature, pH, light and darkness.

Effect of temperature

For the study of temperature requirement in solid media, inoculated Petri plates were incubated at following different temperatures, viz. 1, 5, 10, 15, 20, 25, 20 and 35°C in separate incubators, on the best selected medium.

Effect of hydrogen ion concentration (pH)

The best solid medium was adjusted at different pH levels with Citrate and Phosphate buffers, viz. 3.0, 4.0, 5.5, 6.5, 7.0, 7.5, and 8.0. The inoculated Petri plates were incubated at temperature found to be the best for mycelial growth. The cultures were incubated for 7 days before the growth was measured.

Effect of light and darkness

This was observed only in solid medium. Best selected solid medium with optimum temperature and pH was inoculated and was given light and dark treatment. Growth was observed after 10 days of incubation.

Statistical analysis of cultural data

Cultural data was statistically analysed for Standard Deviation and Critical Difference. Critical Difference was calculated by Analysis of Variance (ANOVA) one-way classification model, as given by Gupta (1985).

ANNEXURE–IV B

Methodology for Qualitative and Quantitative Screening of Enzymes

1) Qualitative screening of vegetative mycelium of morchella *species for extracellular enzymes*

Various ectomycorrhizal fungi are capable of producing extracellular lytic enzymes that also helps them to obtain C, P, and N from various organic compounds. Four species of *Morchella,* viz., *M. angusticeps, M. conica, M. deliciosa* and *M. esculenta* were tested for their ability to produce different extracellular enzymes and were compared also by the qualitative tests described by Hutchinson and Malloch (1988) were used in the present study. Modified Melin-Norkans (MMN) media (Marx, 1969) have been used containing $CaCl_2.2H_2O$ (50 mg), NaCl (25 mg), KH_2PO_4 (500 mg), $(NH_4)_2HPO_4$ (250 mg), $MgSPO_4$ (150 mg), ferric citrate (1.2 ml), malt extract (3.0 mg), glucose (10.0 g), thiamine HCl (1.0 ml), distilled water (1000 ml), agar (0.8%). The pH was adjusted to 5.6 – 5.8.

a) Cellulase test agar

MMN medium containing 0.1% carboxy methyl cellulose (CMC) instead of glucose was prepared and poured into Petri plates. The mycelial discs (3 mm diameter) of *Morchella* spp. were inoculated in each Petri plate and incubated at 28°C for 1-2 weeks. If cellulase enzyme is produced, a clear zone will be formed surrounding the colony, which can be visualized with Congo red dye.

b) Pectinase test agar

MMN medium containing 5 g of pectin/l was used. After inoculation with mycelial disc of respective *Morchella* species, the plates were incubated for 2-3 weeks at 27°C and then these plates were flooded for 12 hours with Congo red dye. If pectinase is produced, a clear zone will occur around the colony, otherwise an opaque background will be observed.

c) Lipase test agar

MMN medium was prepared with 0.1 g of $CaCl_2$ rather than 0.05 g/l. Tween20 was autoclaved separately and added to the medium at a rate of 10 ml/l before pouring into plates. The mycelium of different *Morchella* spp. was inoculated and incubated at 27°C for 2-3 weeks. The breakdown of Tween20 would result in precipitation of $CaCl_2$ crystals in the medium around and below the colony, if the fungus produced lipase.

d) Amylase test agar

MMN medium was prepared by using 2 g/l of soluble starch instead of glucose and poured into plates. The mycelium of *Morchella* species was inoculated and incubated at 27°C for 2-3 weeks. After 4 weeks of incubation, the plates were flooded with Iodine solution (5 g KI and 1.5 g/100 ml of water). The occurrence of a clear zone around the colony indicates the presence of amylase.

e) Urease test agar

MMN medium with 5 g glucose/l, 0.02 g/l phenol red and without ammonium phosphate was prepared in 900 ml of water. Separately, 20 g of urea was added to 100 ml of water and then sterilized. This was added to 900 ml of autoclaved medium to make volume upto 1 litre. 2-3 drops of HCl were added and swirled until medium turned yellow in colour, and then poured into plates. Then the mycelium of *Morchella* species was inoculated in the Petri plates containing medium. If urease enzyme was produced, the medium would turn into red.

2) Quantitative screening of vegetative mycelium of Morchella species for the activity of amylase, protease and cellulase enzymes

For the extracellular production or activity of enzymes, viz., amylase, protease and cellulase, vegetative mycelium of four different species of *Morchella*, viz. *M. angusticeps, M. conica, M. esculenta* and *M. deliciosa* have been used. Macris and Galiotou-Panayotou's (1986) medium was prepared containing 12 g-NaH_2PO_4, 2 g-KH_2PO_4, 7 g-$(NH_4)_2HPO_4$, 0.3 g-$MgSO_4.7H_2O$, 0.3 g- $CaCl_2.2H_2O$, 0.3 g-urea, 0.2 g-protease peptones, 0.1 g-yeast extract, 0.3 ml Tween80, 0.005 g-$FeSO_4.7H_2O$, 0.002 g-$CoCl_2$ and 10 g-starch.

For the specific enzymes, specific enzyme source was used. One per cent starch for amylase enzyme, 1% casein for protease enzyme and 1% cellulose powder for cellulase enzyme was added in the separate flasks having Macris and Galiotou-Panayotou's medium. Double distilled water was added in each flask to make the volume upto 1 litre and the pH 5.0 was adjusted with the help of N/10 NaOH or N/10 HCl. The media were sterilized at 15 psi for 30 minutes.

Three discs of mycelium of *Morchella* spp. were added to 100 ml medium in 250 ml Erlenmeyer flasks. These were incubated at 25°C for 8-10 days and the cultures were then centrifuged at 10,000 rpm for 20 minutes. The supernatant was collected and stored in refrigerator for further assay of respective extracellular enzymes, viz. amylase, protease and cellulase. The activity of the enzymes was expressed in terms of μmole/ml/min (I.U.).

a) Amylase assay

Reference: Reese and Mendel (1963); Miller (1959)

Reagents:

(i) Phosphate buffer (0.1 M, pH 7.0)
(ii) 1% Starch solution (Add 1 gm Starch in 100 ml Phosphate buffer).
(iii) Dinitro salicylic acid (DNSA) reagent: (1 gm DNSA; 1 gm NaOH; 0.05 gm Na_2SO_4; 0.20 gm phenol; 20 gm potassium sodium tartrate were dissolved in double distilled water to make final volume 100 ml).
(iv) Standard maltose (100 μg-700 μg) in double distilled water.

Procedure

To 0.5 ml of starch solution, 0.3 ml of phosphate buffer (0.1 M, pH 7.0) and 0.2 ml of enzyme solution were added. The control without enzyme was also run. Contents were incubated at 55°C for 5 minutes, cooled to room temperature and 3 ml of DNSA reagent was added to each tube. 0.2 ml of enzyme solution was added to the control tube and heated in boiling water bath for 30 minutes. After cooling to room temperature, absorbance (OD) at 540 nm was measured against a reagent blank.

Enzyme activity

One unit of amylase activity was defined as the amount of enzyme that released 1 mole of maltose from starch/minute under the assay conditions. Standard curve was prepared from 1 μ mole to 100 μ mole of maltose.

b) Protease assay

Reference: Manachini *et al.* (1988)

Reagents:

(i) 50 mM Tris-HCl in 2 mM $CaCl_2$ (pH 7.0)

(ii) 0.625% Casein (Analytical Grade)

(iii) 5% Trichloroacetic acid

(iv) Standard tyrosine (1 μ mole-100 μ mole) in double distilled water.

Procedure

To 4 ml of casein solution, 1 ml of enzyme solution was added and incubated at 55°C for 15 minutes. The reaction was terminated by adding 5 ml of 5% trichloroacetic acid and the precipitates formed were removed by filtration using Whatmann filter paper No. 1. Absorbance (OD) was measured at 275 nm against reagent blank containing 1 ml double distilled water.

Enzyme activity

One unit of proteolytic activity was defined as the amount of enzyme that released 1 μ mole of tyrosine from casein/minutes under assay conditions. Standard curve was prepared from 1 μ mole to 100 μmole of tyrosine, following the same procedure.

c) Cellulase CMCase assay

Reference: Reese and Mendel (1963); Miller (1959)

Reagents:

(i) Citrate buffer (0.055 M, pH 4.0)

(ii) Carboxymethylcellulose solution: 1% Carboxymethylcellulose (CMC) in citrate buffer (0.055 M, pH 4.0)

(iii) Dinitro salicylic acid (DNSA) reagent: (1gm DNSA; 1 gm NaOH; 0.05 gm Na_2SO_4; 0.20 gm Phenol; 20 gm potassium sodium tartrate were dissolved in double distilled water to make final volume 100 ml).

(iv) Standard D-glucose (20 μg-200 μg) in double distilled water.

Procedure

To 0.5 ml of enzyme preparation, 4.5 ml of carboxymethylcellulose solution [1% Carboxymethylcellulose (CMC) in citrate buffer (0.055 M, pH 4.0)] was added. The control was also run with all components except enzyme. These were incubated at 55°C for 15 minutes, cooled to room temperature and 0.5 ml of enzyme preparation was added to the control. One ml of mixture was withdrawn from the reaction mixture and 3 ml of DNSA reagent was added to each tube. The mixture was heated in boiling water bath for 45 minutes. After cooling to room temperature, absorbance (OD) was measured at 540 nm against a reagent blank.

Enzyme activity

One unit of cellulase activity was defined as the amount of enzyme that released 1 μ mole of D-glucose from CMC/minute under assay conditions. Concentration of reducing sugars was calculated from a standard curve of D-glucose (20 μg-200 μg/ml).

ANNEXURE–V

Methodology for Determination of Nutritional Components

The carpophores of *Morchella* species collected during the fruiting season were chemically analyzed for various components, e.g. nitrogen, proteins, amino acids. Total free amino acids, starch, carbohydrates, reducing and non-reducing sugars, and macro and micro- nutrients were estimated following the standard techniques outlined below:

(i) Determination of nitrogen and crude proteins

Nitrogen was estimated by Micro-Kjeldhal method (Jackson, 1958). Dried samples (250 mg) were taken in Micro-Kjeldhal flasks. To these were added 2 gm mixture (K_2SO_4 – 400 gm + $CuSO_4$ – 20 gm + HgO – 3 gm + Selenium powder – 1 gm) and 15 cc of sulphuric acid. Flasks were left for digestion till the solutions of flasks were clear white. Final volume was made with 200 ml of distilled water or deionized water. After adding 50 cc sodium hydroxide (40%), it was distilled in 300 ml flasks. Ammonia gas was trapped in 250 ml conical flaks with 10 cc boric acid (4%) and 2 drops methylene blue indicator, and titrated with N/10 H_2SO_4. Readings were compared with chart for equivalent percentage of nitrogen.

The total crude proteins were estimated by multiplying the per cent nitrogen content with factor 6.25 (Crisan and Sands, 1978).

(ii) Estimation of other nutrients (P, K, Mg, Ca, Si, Zn, Cu, Mn, Fe)

A known weight of the sample was digested in a diacid mixture of HNO_3 and $HClO_4$ in a ratio of 4:1 v/v and the digestion was made to a specific volume. Phosphorus was estimated by Vanadomolybdo Phosphoric Yellow Colour method (Jackson, 1967) and potassium by Flame Photometer method, respectively. Calcium, magnesium, iron, manganese, copper and zinc was determined on Pye Unicam Sp – 1900 Atomic Absorption spectrophotometer.

(iiii) Determination of total sugars

Total sugars were determined by the method of Dubois *et al.* (1956). The extraction was done by homogenizing 100 mg of the dried sample in 2 ml of 30 per cent ethanol. Subsequently, centrifugation was done at 500 g for 10 minutes. The supernatant was collected and the residue was again extracted with 2 ml of 80 per cent ethanol followed by centrifugation at 500 g for 10 minutes. The supernatants were pooled and then reduced to dryness by heating 60°C. The final volume was made to 2 ml with 80 per cent ethanol, and 0.1 ml of each solution was taken for estimation.

Reagents:

The following reagents were prepared:

(a) Phenol reagents by mixing 10 ml of distilled water with 90 ml of 90 per cent phenol solution (90 g of phenol dissolved in distilled water to make the final volume – 100 ml)

(b) Conc. sulphuric acid.

(c) Standard glucose (25 to 250 μg/ml) solution.

(d) Extract (0.1 ml) was raised to 2 ml with distilled water. To this was added 0.05 ml of phenol reagent and thoroughly mixed. Then 5 ml of conc. sulphuric acid was rapidly added and the mixture was allowed to stand at room temperature for 30 minutes. The absorbance was read at 485 nm. Standard curve was prepared by taking glucose as the standard sugar solution and the value of total sugar was calculated from this reference curve.

(iv) Determination of starch

The starch content was determined by method of McCready *et al.* (1950). For extraction, 100 gm of the dried sample was homogenized in 80 per cent ethanol. The homogenate was centrifuged at 500 g for 10 minutes and residue was washed repeatedly with 80 per cent ethanol to remove all traces of soluble starch. Every tine supernatant was pooled together and used for the assay of total sugars. The residue was freed of alcohol by adding 3 ml of distilled water in the test tube and boiled for 30 minutes. The alcohol free residue was used for estimation of starch.

To the residue, 0.1 ml of distilled water was added and kept at 70°C for half an hour. After that 1.0 ml of 52 per cent prechloric acid was added and centrifuged at 3000 g for 15 min. The supernatant was collected.

Anthrone reagent

Anthrone–sulphuric acid reagent was prepared by dissolving 200 mg of anthrone in 100 ml of cold H_2SO_4.

Estimation

To 0.2 ml of extract, 0.8 ml of distilled water was added, and then 5 ml of cold Anthrone-sulphuric acid reagent was added with constant shaking. The reaction mixture was heated for 8 -10 min in boiling water bath and cooled rapidly to room temperature. Optical density was recorded at 630 nm. Standard curve was prepared by using glucose concentrations in the range 250 to 500 μg/ml.

(v) Determination of reducing sugars

The reducing sugars were estimated by dinitro salicylic acid (DNSA) method given by Summer (1935). The supernatant was collected in china dish and evaporated to dryness on a hot water bath. The residue was eluded with 2 ml of 80 per cent ethanol.

DNSA (2.5 g) was mixed with 50 ml of distilled water containing 4 g of NaOH. Then sodium potassium tartrate (75 g) was added to it and total volume was made to 250 ml with distilled water.

Estimation

To 0.5 ml of extract, 0.5 ml of distilled water and 1.0 ml of DNSA was added in each sample.

Reaction mixture was boiled for 10 min, cooled and diluted with 2 ml of distilled water. The optical densities were recorded at 560 nm. The quality of reducing sugars was calculated against the reference curve prepared by using glucose.

(vi) Determination of total free amino acids

The total free amino acids were determined by the method of Lee and Takahashi (1966). 10 mg of the dry material of each sample was extracted in 2 ml of 80 per cent and centrifuged at 500 g for 10 minutes. The extraction was again repeated with 2 ml of 80 per cent ethanol, centrifuged at 500 g for 10 minutes so as to allow complete removal of amino acids from the sample. Pooled the supernatants and reduced them to dryness at 60°C. The final volume of the solution was made to 2 ml with 80 per cent ethanol.

Reagents:

(a) 0.5 M citrate buffer pH 5.5.

(b) Ninhydrin solution: One per cent ninhydrin in 0.5 m citrate buffer.

(c) Glycerol solution: 100 per cent glycerol was used as such.

(d) Reaction mixture of Ninhydrin-citrate: 1 per cent ninhydrin solution (in 0.5 M citrate buffer) was mixed with glycerol: Citrate buffer (5:12.2) and the pH of the solution was adjusted to 6.0 with the help of sodium citrate.

(e) Glycine standard solution (10 to 100 μg/ml) 100 mg. Glycine dissolved in 10 ml of distilled water.

To 0.2 ml of the extract, 2 ml of ninhydrin, citrate and glycerol mixture was added. The contents were heated in a boiling water bath at 100°C for 12 minutes and then cooled to room temperature. The absorbance was read at 570 nm. Quantity of total free amino acids was calculated from standard curve prepared with glycine.

(vii) Chromatographic analysis of free amino acid

Unidirectional (ascending) Chromatographic method using Whatman No. 1 Chromatographic paper was followed. 0.1 ml of extract, made for estimation of total free amino acid, was applied on chromatographic sheet and Butanol : Acetic acid : Water (3:2:5) was employed as a solvent system. The paper was removed when the stimulant had reached 2 cm less than its upper surface. The chromatographic paper was dried at room temperature for 24 hours and sprayed with 0.2 per cent ninhydrin in butanol and dried again at room temperature. The chromatogram was activated at 80°C in an oven for 20 minutes and their Rf values for the different amino acids were calculated and compared with standard amino acid Rf values.

(viii) Determination of moisture content

The sporophores were weighed immediately after collecting and dried for a minimum of 3 days (Oi-Wah Lau, 1982) in an oven at 105 ±° C to record constant weight. The total solid material left after the loss of moisture was recorded.

ANNEXURE–VI

Methodology for Determination of Nutraceutical Potential of Morels

Sample preparation

The fruiting bodies of *Morchella* species collected for analysis of nutraceuticals were sun dried and stored in brown paper bags. They were analyzed for various nutraceutical components, viz. protein, fibre, amino acids, vitamins, cholesterol level, and minerals following the standard techniques as outlined below. The analysis of these components was done at Choksi Laboratories Limited, Indore, Madhya Pradesh - 452 001.

1. Determination of moisture content (A.O.A.C., 1980)

Oven drying method

Principle: This method measures the weight loss in foods due to the evaporation of water. Moisture content can be determined by loss of weight after thorough drying. Drying methods are generally used as they give accurate results. Use an oven whose temperature can be controlled accurately. Since the temperature tends to be different on different shelves, a drying oven fitted with an internal fan for circulation of air is preferred.

Procedure: A flat bottom metallic dish was taken and weighed. 5.1250 g of the sample was spread over it. Now the dish was kept in a hot air oven at 110°C for about one hour. After that the disc was taken outside and covered properly. Now the disc was kept in a desiccator. After 10 minutes the weight of the dish was noted. The difference in weight after and before drying the samples gives the moisture content of the sample and is expressed as the total percentage of moisture content.

2. Determination of ash content (A.O.A.C., 1980)

Principle: Ash content of foodstuffs represents the inorganic residue remaining after destruction of organic matter. It may not necessarily be exactly equivalent to the mineral matter as some losses may occur due to volatilization or some due to interaction between the constituents. High ash content and/or a low alkalinity of the ash may in some cases be suggestive of the presence of adulterants. Ash content in general is the weight of the mushroom after ignition to constant weight in a furnace at about 550°C.

Procedure: Weigh three dry silica dishes (7-8 cm diameter). Weigh about 5-10 g of the mushroom sample into silica dishes and keep them in a muffle furnace. Ignite the dish and the contents at 525–550°C for 4 to 6 hrs until the sample turned into ash (if needed, ash overnight). Cool down the dishes in desiccator and weigh. The difference in weights before and after drying gives the total ash content and is expressed in the terms of the total percentage of ash in the respective mushroom species.

3. Estimation of crude fat content (A.O.A.C., 1980)

Principle: Fat-soluble material in a food is extracted from an oven dried sample using a Soxlets Extraction Apparatus. Ether is evaporated and the residue is weighed.

Apparatus: Soxlets Extraction Apparatus.

Reagents: Anhydrous/analytic reagent grade ether.

Procedure: About 10 g of the sample was dried in oven at 100°C for about 1 hour. 8.585 g of sample was taken and extracted with ether using Soxlets extraction apparatus. Then ether was evaporated at about 80°C and residue was weighed. The total crude fat is expressed as the total percentage.

$$\% \text{ Crude fat} = \frac{\text{Weight of ether extract (g)}}{\text{Weight of the sample (g)}} \times 100$$

4. *Estimation of crude fibre (A.O.A.C., 1980)*

Principle: Crude fibre is the organic residue that remains after the food sample has been treated under standardized condition with petroleum sprit, boiling dilute sulphuric acid, boiling dilute sodium hydroxide solution and alcohol. The mushroom is first made fat free and is then digested successively with acid and base, and the crude fibre is deduced from the loss in weight on ignition of dried residue remaining after the digestion. Care should be taken to avoid fibre contamination from paper or brush.

Reagents: Sulphuric acid solution (1.25%) – 0.7 ml of conc. H_2SO_4 per 100 ml.

Procedure: Two g of dry material (mushroom) extracted with ether or use the residue from crude fat determination. 200 ml of the boiling sulphuric acid solution is added immediately, and the digestion flask is connected with condenser and then heated. After 30 minutes of boiling, remove the flask and immediately filter through filtering cloth in afluted funnel.

$$\% \text{ Crude fibre} = \frac{\text{Loss of weight noted}}{\text{Weight of sample}} \times 100$$

5. *Determination of protein content*

(Micro-Kjeldahl Distillation Method) – Jackson (1958)

Principle: Total protein content, determined by the Micro-Kjeldahl method, is defined as the amount of nitrogen experimentally found and multiplied by an appropriate conversion factor. The sample is oxidized in the presence of sulphuric acid and nitrogenous compound is converted into ammonium sulphate. Mercury is added to the digestion mixture as a catalyst and alkali sulphate as a boiling point elevator. Ammonia is liberated by adding an excess of alkali and is quantitatively distilled into a measured volume of standard HCl or H_2SO_4 acid.

Procedure: 1.7544 g of sample is taken and is digested with 40 ml of conc. H_2SO_4 acid. The entire nitrogenous compound is converted into ammonium sulphate.

$$\text{Organic Nitrogen} + H_2SO_4 \rightarrow (NH_4)_2\, SO_4 + H_2O + CO_2 + \text{other by-products}$$

Mercuric oxide was added as a catalyst, as mercury forms a complex with ammonium ions during digestion. Sodium sulphate or sodium thiosulphate was added to break the complex and mercury sulphide was added (it also elevates boiling point). The liberated ammonium gas is collected in standardized hydrochloric acid. The excess acid was neutralized with sodium hydroxide (standardized).

The total crude protein content was estimated by multiplying the per cent nitrogen content with factor 6.25 (Crisan and Sands, 1978).

Calculation for N_2 Determination:

Weight of sample taken: x gm

Volume of 0.1N HCl used : y ml

Normality of 0.1 N HCl : Z

Percentage N_2 Content:

Total Crude Protein: .... × 6.25

= ....% w/w

6. *Estimation of metals*

(i) Ashing and Digestion Process (for Se, Fe, Ca, Zn, Mn, and Cu):

1. About 15 g of sample was taken in a silica crucible.
2. Sample was charred and then kept in muffle furnace for ashing at 450-500°C.
3. After complete ashing, crucible was allowed to cool.
4. Sample was digested with conc. HCl and conc. HNO_3 (1-2 ml each).
5. Sample was filtered and made up to 25 ml volume and aspirated on Atomic Absorption Spectrophotometer.

(ii) Ashing and Digestion Process (for Na and K):

1. About 15 g of sample was taken in a silica crucible.
2. Sample was charred and then kept in muffle furnace for ashing at 450-500°C.
3. After complete ashing, crucible was allowed to cool.
4. Sample was digested with conc. HCl and conc. HNO_3 (1-2 ml each).
5. Sample was filtered and made up to 25 ml volume and reading was taken on Flame Photometer.

7. *Determination of the energy value*

The energy value of the mushrooms can be determined by the content of the crude protein (N × 6.25), fat and carbohydrates in mushroom using the factors 2.62, 8.37 and 4.2 Kcal/g of each component, respectively (Crisan and Sands, 1978).

Energy value (in Kcal/100 g dry weight) =2.62 × (%N × 6.25) + 8.37 × (% fat) + 4.2 × (% carbohydrates)

HPLC (High Pressure (performance) Liquid Chromatography) Analysis

Principle: It separates the liquid sample into the components. The sample is injected into a stream of high pressure liquid (mobile phase), which sweeps the sample into column (stationary phase) and then to detectors. Separation takes place in column as different components will have a different chemical affinity for the column packing. Those with less affinity will exist first (lesser elution time). Each component elutes from the column as peak. The time that a peak takes to exit the column determines the component species. It is called retention time. Peak height and peak area determine the quantity. Different components have different rate of adsorption on stationary phase, which depends upon the polarity and chemical nature of solute. Mobile phase sweeps away the less absorbed solute, and it is detected by the detector firstly. Area under the curve determines the quantity of solute and retention time determines the identity of solute.

Calculations:

$$\frac{\text{Area of Test Peak}}{\text{Area of Standard Peak}} \times \frac{\text{Weight of Standard}}{\text{Dilution}} \times \frac{\text{Dilution}}{\text{Weight of Test}} \times \frac{\text{Polarity of Standard}}{100}$$

1. *Estimation of nucleic acids*

Standard Preparation: Prepare 100 ppm standard solution in mobile phase.

Sample Preparation: Ten g of the sample was extracted with mobile phase and 20 µl was injected into HPLC system.

HPLC Conditions:

Column	:	Gemini C-18, 5 micron, 250 × 4.6 mm
Detector	:	UV
Wavelength	:	254 nm
Flow	:	1 ml/min.
Oven temperature	:	Ambient
Mobile phase	:	0.025 M KH_2PO_4 (pH 3):Acetonitrile (98:2)
Injection volume	:	20 µl

2. *Estimation of vitamins*

a) Water-soluble vitamins:

Principle: Water-soluble vitamins (B and C) are dissolved in aqueous acetic acid and their ionic strength is adjusted. They are separated by paired ion high performance liquid chromatography and quantified by comparing the absorbance of standard with sample.

Standard Preparation: Prepare 1000 ppm stock solution in mobile phase A. Pipette out 10 ml of each into 100 ml volumetric flask and add 1 gm $CaCO_3$. About 25 ml of mobile phase A was added, cooled and then diluted to 100 ml with mobile phase A.

Sample Preparation: Ten gm of sample is extracted with mobile phase on magnetic stirrer and 50 µl is injected into HPLC system.

HPLC Conditions:

Column	:	Gemini C-18, 5 micron, 250 × 4.6 mm
Detector	:	UV
Wavelength	:	286 nm
Flow	:	1 ml/min.
Oven temperature	:	Ambient
Mobile phase	:	Mobile phase A: Mobile phase B
Injection volume	:	50 µl

Part A: In 2 litre volumetric flask containing 1000 ml water, dissolve 2.0 g pentane sulphonic acid sodium salt and 20 ml acetic acid. Dilute to the volume with water.

Part B: In 2 litre volumetric flask containing 1000 ml methanol, dissolve 20 g pentane sulphonic acid sodium salt and 20 ml acetic acid. Dilute to volume with methanol.

Final composition (20% B and 80% A)

(b) Fat-soluble vitamins:

Standard preparation: Prepare standard stock solution of vitamin A (1000 ppm) in isopropyl alcohol; vitamin D and E (1000 ppm) is prepared in methanol.

Sample preparation: Take 1 ml of each solution in 10 ml flask and dilute to the volume with methanol.

Isolation of unsaponifiable matter: Accurately weigh 50 g sample into saponification flask. Add 100 ml alcohol, 25 ml, 25 % aqueous sodium ascorbate solution, and 25 ml 50% (w/v) aqueous KOH solution. Reflux for 45 minutes on steam bath. Cool down rapidly under running water. Transfer the liquid to separator with 2, 75 ml portion water, 2, 25 ml portion of alcohol and 2, 100 ml portion of ether. Shake vigorously for one minute and let that stand until both layers are separated and clearly distinguished. Transfer the aqueous phase to second separator and shake with mixture of 25 ml alcohol and 100 ml of pentane. Let it separate, and transfer the aqueous phase to third separator and pentane phase to first separator, washing second separator with 2, 10 ml portion pentane and adding washing to first separator. Wash out the combined pentane extract with 3, 100 ml portion freshly prepared 3% solution of KOH in 10% alcohol, shaking vigorously. Then washed with 100 ml portion Water until last washing is neutral to phenolphthalein. Pass through sodium phosphate.

Transfer dried pentane test solution to 500 ml round bottom flask, rinsing separator and paper with pentane and add 1 ml of antioxidant solution. Evaporate to dryness under vacuum by swirling in water bath at 40°C. Cool under running water and restore atmospheric pressure with N_2. Dissolve the residue immediately in 5 ml methanol.

HPLC Conditions:

Column	:	ODS 250 mm × 4.6 micron
Detector	:	UV
Wavelength	:	280 nm
Flow rate	:	1.5 ml/min
Column oven	:	25°C
Mobile Phase	:	Methanol 100%
Injection volume	:	20 μl

Injected 20 μl into chromatographic system and the samples were run for 20 minutes.

3. Estimation of amino acids

Principle: Amino acids are derivatized with orthophalaldehyde, and the compound which forme produces fluorescence, and are detected by HPLC using fluorescence detector.

Standard preparation: Prepare 100 ppm stock solution in mobile phase. Mix 1:1 stock solution and Ophthalaldehyde solution in methanol and inject 20 μl.

Sample preparation: Ten g sample was extracted with mobile phase and derivatized and injected

HPLC Conditions:

Column	:	Gemini C-18, 5 micron, 250 × 4.6 mm
Detector	:	Fluorescence detector
Wavelength	:	254 nm
Flow	:	1 ml/min.
Oven Temperature	:	Ambient
Mobile phase	:	CAN: 25 mM Potassium Phosphate (pH 3.3) 80:20
Injection Volume	:	50 μl

4. Estimation of Polysaccharides

Standard preparation: Prepare all saccharides (100 mg) in 10 ml of water, and sonicate for 20 minutes.

Sample preparation: Two g of the mushroom sample was dissolved in 50 ml of 0.1N HCl and then sonicated for 20 minutes.

HPLC Conditions:

Column	:	Amino (waters) 250 m × 4.6 mm
Detector	:	Refractive index detector
Flow	:	2 ml/min.
Mobile phase	:	Acetonitrile : water (85 : 15)
Injection Volume	:	20 μl

Gas Liquid Chromatography Analysis

Principle: Gas chromatography, specifically gas liquid chromatography, involves a sample being vaporized and injected onto the head of chromatographic column. The sample is transported through column by the flow of inert gaseous mobile phase (carrier gas). The column itself contains a liquid stationary phase, which is adsorbed onto the surface of an inert solid. When a vaporized compound is introduced into a carrier gas and carried into the column, it is partitioned between gas and stationary phase by a dynamic counter current distribution process. The compound is carried down the column by the carrier gas, returned to a greater or lesser extent by sorbtion and desorbtion on the stationary phase. The elution of the compound is characterized by that portion ratio.

1. Estimation of Cholesterol Level

Principle: Lipid is extracted from sample by mixing solvent and saponified. Unsaponifiable fraction contains cholesterol & other sterols and is extracted with benzene. Sterols are derivatized to form Triethylsilyl (TMS) ethers, which are detected quantitatively by GLC (Gas Liquid Chromatography.

Standard preparation: 100 ppm

Sample preparation: Lipid is extracted from 8.5851 g. of sample (from which 2.78% fat is obtained) by mixed solvent and saponified. Unsaponifiable fraction contains cholesterol and other sterols are extracted with benzene. Sterols are derivatized to form triethylsilyl (TMS) ethers, which are detected quantitatively by GLC.

GC Conditions:

Column	:	OV-17
Detector	:	FID
Flow	:	30 ml/min.
Oven temperature	:	235°C
Injector temperature	:	250°C
Detector temperature	:	250°C

ANNEXURE–VII

Methodology for Rhizomorphosphere Studies

Soil samples for mycofloral studies of *Morchella* species were brought from different localities in Himachal Pradesh. These sites were located in areas of maximum morel production. The morel bearing soils for all the species investigated, from different forest types and the component trees, were collected in triplicate. The samples were collected in sterilized polythene bags. Only the soil associated with the rhizomorph was collected.

i) Isolation of fungi

For the isolation of fungi, the Dilution plate method (Warcup, 1960) was employed. In this method, one gm of soil is dissolved in 1000 ml of distilled water. From this 10 ml is taken and added to 100 ml of distilled water. Then 1 ml solution was taken and dispersed over medium. Isolation from these dilutions were made on Czapek's Dox (Raper and Thom, 1949) medium. The Petri plates were sterilized in autoclave at 15 lbs pressure for 15 minutes. The sterilized medium was poured into the Petri plates under aseptic conditions (pre-sterilized UV chamber) and allowed to solidify before isolations were carried out. Samples of each category were taken with the help of a pipette and poured into Petri dishes containing medium, which were slightly tilted, so that the dilution could spread evenly on the medium. The inoculated Petri plates were incubated at 22 ± 2°C.

ii) Microscopic examination

The Petri plates were examined regularly for the growth of fungi. Temporary mounts of fungi were made in cotton blue* and Lactophenol**. The slides were examined under a Getner Trinocular Microscope, having a magnification of x 1040 under 10 x 45 lenses.

a Preparation of pure cultures

For obtaining pure cultures of fungi, Czapek's Dox medium was poured into test tubes and sterilized at 15 lbs pressure for 15 minutes and made into slants. Subculturing of different fungi appearing in Petri plates was done into the slants, which were then incubated at 22 ± 2°C.

b Identification

The cultures were identified following Barnett and Hunter (1972) and Gilman (1967). Identifications were authenticated by Prof. K.G. Mukerji, Department of Botany, University of Delhi.

***Cotton blue**	-	1.0 gm		
Lactophenol	-	1.0 ml		
****Lactophenol**:		Lactic acid	-	25.0 ml
		Phenol	-	25.0 ml
		Glycerin	-	25.0 ml
		Distilled water	-	25.0 ml

iii Statistical analysis

The four forest types and the component trees were compared for each species using Sorenson's Index of similarity (1948), which has the formula S = 2 c / a + b x 100, where 'a' is the total number of species in one forest type, 'b' is the total number of species in the second forest type, and 'c' is the number of species common to both of the forest types. The index ranges from zero, if the two types have no species in common and to 100, if all species are shared in common. This was also done for six species of *Morchella*.

ANNEXURE–VIII

Methodology for Sociobiological, Ethnomycological and Culinary Studies

Indigenous knowledge is the integrational wisdom of local inhabitants to perform their livelihood operations. Indigenous people (rural communities/tribals/ethnic societies) have invaluable knowledge, which is transferred orally from one generation to another. The ethnomycological knowledge about consumption, medicinal uses, and other facts related to the biology of morels has been passed from one generation to another orally. This Indigenous Technical Knowledge (ITK) of *Morchella* species was gathered by PRA technique.

(PRA) Technique and Field study

The technique followed to gather information was **Participatory Rural Appraisal (PRA)**. For gathering information, regular forays were conducted in different parts of the tribal belt during the fruiting period of every year. List of the places/villages visited are given in the study area. The data was gathered in the villages during social gatherings of the local inhabitants mostly in 'gram sabhas', 'mahila mandals' and 'yuvuk mandals' or 'jan mandals'. The gatherings were mostly at a common place of their local deity. Every village has its own deity, so the people gather at that common place to discuss their day to day issues. Such meetings were conducted in the evenings when every body was free from daily routines.

For collecting data on ethnomycology, the villagers in such gatherings were interviewed to obtain information on various ethnomycological aspects such as collection, appearances, preservation, consumption, marketing, and ethnic use in food and medicine, etc. All the information regarding 'spring morels' and 'monsoon morels' were recorded on a specially designed questionnaire.

Pro forma for recording data on "Ethnic use of *Morchella* in food and medicine in Tribal areas of Himachal Pradesh"

A. General Information:

1. Name of the area ....Tehsil ....
   Village ....

B. Information about the collection:

1. Specimens collected ....
2. Local Name ....
3. Habit ....
   I. Habitat ....
   II. Association ....
   III. Relevant ecological data:
      a. Forest Type ....
      b. Altitude ....
      c. Collection No. ....
      d. Season (winter/rainy) ....

C. Specific Information:

1. Name of the Household ....
2. Number of family members ....
3. Family members engaged in collection of *Morchella* ....
4. Whether holding Licence, Yes / No ....
5. Quantity collected by each member during the season ....
6. Total quantity collected ....
7. Quantity used for self-consumption ....
8. Quantity marketed ....
9. Rate/Kg ....
10. (Marketing) Value
    a. Spring morels ....
    b. Monsoon morels ....
11. Mode of disposal direct/ through middleman ....
12. Total annual income of the household ....

D. Miscellaneous Information:

1. Myths and local-beliefs associated with occurrence ....
   ....
   ....

2. Frequency of occurrence ....
....
3. Impact of vegetation:
    a. On occurrence ....
    b. Overall colour of the species ....
    c. Colour of the ridges ....
4. Local recipes ....
....
....
....
....
5. Drying of specimens ....
....
6. Folks ....
7. The use of mushroom ....
8. The consumption of the mushroom and its effect on the health of the people ....
....
....
....
9. The various recipes used and ingredients ....
....
....
10. The medicinal value ....
....
....
....
....
....

ANNEXURE–IX

Methodology for Mycorrhizal Studies

During an extensive survey conducted on morels and other mushrooms of N-W Himalayas, three species of *Morchella* have been found to enter into mycorrhizal relationship with different plant species. Mycorrhizal connections were established by digging the soil and tracing the roots of associated plants to the fructification following Young (1936, 1940) and Zak (1971, 1973). The roots of plants associated were thoroughly washed with tap water. After fixing them in FAA* for 24 hours, they were preserved in 70% alcohol. In order to confirm the formation of mycorrhizae, both stained (cotton blue in Lactophenol) and unstained sections were prepared and examined microscopically at various magnifications with the help of a research microscope (Getner Binocular Research Microscope GTRbA) 10x, 40x, and 100x magnifications. All the microscopic structures were drawn with the help of Camera Lucida.

---

*FAA

Ethyl Alcohol 70%	-	90 ml
Glacial Acetic acid	-	05 ml
Formaldehyde 40%	-	05 ml

ANNEXURE-X

Methodology for Cytological Studies

Collection of Material

The fruiting bodies of *Morchella esculenta, M. conica, M. angusticeps, M. deliciosa* and *M. crassipes* were collected at various stages of development from different forest types.

Fixation: The fruiting bodies collected were fixed in freshly prepared BAC fixative (Lu, 1962) and Acetic Alcohol fixatives for 24 hours in glass tubes. After 24 hours the fruiting bodies were transferred to 70 per cent alcohol.

Staining: The tissues from the pits were stained in Propiono-Iron-Haematoxylin stain (Henderson and Lu, 1968).

(i) Preparation of stain

Propiono-Iron-Haematoxylin stain consists of two solutions. Solution A is 0.5 per cent Iron alum (Ferric ammonium sulphate dissolved in 50 per cent propionic acid, to which few drops of saturated solution of ferric acetate were added). Solution B is 2 per cent haematoxylin in 50 per cent propionic acid. Solution A and B is then mixed in the ratio of 1:2 and kept in refrigerator for 24 hours.

(ii) Preparation of Slides

Tissue from the pits of the fruiting bodies was squashed in this mixture, coverslip was placed on material and the slide was heated gently. The nuclei stained immediately.

Temporary squash preparations were made permanent by passing them through Tertiary Butyl Alcohol and Glacial Acetic acid mixture (1:3) and then Tertiary Butyl Alcohol (Ranga Rao and Mukerji, 1971) and mounted in Canada balsam. Slides were observed with the help of a research microscope (Getner Binocular Research Microscope GTR 6A). The magnification of the system with various combinations of eye piece and objective lenses was as: x 2350 (with 10 x 100), x 1040 (10 x 45) and x 235 at (10 x 10).

All the microscopic structures were drawn with the help of Camera Lucida. Photographs were taken using 120 SA film.

References

A.O.A.C. 1980. Official Methods of Analysis. (13th ed.). Association of Official Analytical Chemists. Washington DC, USA.

Aaronson, S. 2000. Fungi. *In*: The Cambridge world history of food (Eds. K.F. Kiple and K.C. Ornelas). Cambridge University Press, pp. 958.

Adachi, K., Nanba, H. and Kuroda, H. 1994. The effect enhancement of cytokine production by macrophages stimulated with 1, 3 beta D glucan, grifolan, isolated from *Grifola frondosa*. Biol. Pharm. Bull. **17**: 1554-1560.

Adhikari, M.K. 1995. Toxic and Medicinal Mushrooms from Nepal. Network (Newsletter of the National Committee for APINMAP/SCAMAP, Nepal), **1**: 1-2.

Adhikari, M.K. 1999. Wild relatives of some arable mushrooms found in Nepal. *In*: National Conference on Wild Relatives of Cultivated Plants in Nepal. Green Energy Mission, Kathmandu, pp. 149-155.

Adriano, F.T. and Cruz, R.A. 1933. The Chemical Composition of Philippine Mushrooms. Phillip. J. Agric. **4**: 1-11.

Ainsworth, G.C. 2001. Ainsworth and Bisby's Dictionary of Fungi. 9th ed. CABI Kew, Surrey, England, pp. 650.

Akpaja, E.O., Isihuemhen, O.S. and Okhuoya, J.A. 2003. Ethnomycology and usage of edible medicinal mushrooms among the Igbo people of Nigeria. Intern. J. Med. Mush. **5**: 1-12.

Alcron, J. 1984. Hvastec Mayan Ethnobotany. Univ. Texas, pp. 982.

Allen, L.V. 1997. Nutritional Products. *In*: Handbook of Non-prescription Drugs (Eds. T.R. Covington, R.R. Berardi, L.L. Young, S.C. Kendall, and M.J. Hickey) Washington, DC: American Pharmaceutical Association. pp. 280-287.

Amir, R., Levanon, D., Hadar, Y. and Chet, I. 1992. Formation of Sclerotia by *Morchella esculenta*: relationship between media composition and turgor potential in the mycelium. Mycol. Res. **96**: 943-948.

Amir, R., Levanon, D., Hadar, Y. and Chet, I. 1993. Morphology and Physiology of *Morchella esculenta* during sclerotial formation. Mycol. Res. **97**: 683-689.

Amir, R., Levanon, D., Hadar, Y. and Chet, I. 1994. The role of source-sink relationship in translocation during sclerotial formation by *Morchella esculenta*. Mycol. Res. **98**: 1409-1414.

Amir, R., Levanon, D., Hadar, Y. and Chet, I. 1995a. Factors affecting translocation and sclerotial formation in *Morchella esculenta*. Exper. Myco. **19**: 129-136.

Amir, R., Levanon, D., Hardar, Y. and Chet, I. 1992. Formation of sclerotia by *Morchella esculenta*: Relationship between media composition and turgor potential in the mycelium. Mycol. Res. **96**: 943:48.

Amir, R., Steude, E., Levanon, D., Hadar, Y. and Chet, I. 1995b. Turgor changes in *Morchella esculenta* during translocation and sclerotial formation. Exper. Myco. **19**: 129-136.

Anderson, E.E. and Fellers, C.R. 1942. The food value of mushrooms (*Agaricus campestris*). Proc. Am. Soc. Hort. Sci. **41**: 301-304.

Andlauer, W. and Furst, P. 2002. Nutraceuticals: a role piece of history, present status and outlook. Food Res. Intern. **35**:171-176.

Arai, S. 1996. Studies on functional foods in Japan – State of art. Biosci. Biotech. Biochem. **60**: 9-15.

Arora, D. 1991. All that the Rain Promises, and more............". A Hip pocket guide to Western Mushrooms. Ten Speed Press, California, pp. 263.

Atkinson, G.F. 1961. Studies of American fungi. Hafner Publishing Company. New York, pp. 322.

Ausubel, F.M., Brent, R., Kingston, R.E., Moore, D.D., Seidman, J.G., Smith, J.A. and Struhl, K. 1995. *In:* Current Protocols in Molecular Biology Vol. 1. John Wiley & Sons, Inc.

Babineau, T.J., Kim, D.H. and Han, M.J. 1994a. A phase II multi centre double-blind randomized placebo-controlled study of three dosages of an immunomodulator (PGG-glucan) in high-risk surgical patients. Arch. Surg. **129**: 1204-1210.

Babineau, T.J., Kim, D.H. and Han, M.J. 1994b. Randomized phase I/II trial of a macrophage-specific immunomodulator (PGG-glucan) in high risk surgical patients. Ann. Surg. **129**: 1204-1210.

Baker, K.F. and Cook, J.R. 1974. Biological Control of Plant Pathogens. W.H. Freeman, San Francisco, pp. 433.

Baker, K.F. and Matkin, O.A. 1959. An unusual occurrence of morels in cultivated beds of *Cymbidiums*. Plant Dis. Reptr. **43**: 1032.

Bakshi, B.K. 1974. Mycorrhiza and its role in Forestry. F.R.I. Dehradun, PL – 480, Project Report. pp. 89.

Bano, Z. 1976. The nutritive value of mushrooms. *In*: Proceedings of first symposium on Survey and Cultivation of Edible Mushrooms of India. Regional Res. Lab. Srinagar, **2**: 148-160.

Bano, Z. and Rajarathnam, S. 1982. *Pleurotus* mushroom as nutritious food. *In*: Tropical Mushroom-Biological Nature and Cultivation Methods (Eds. S.T. Chang and T.H. Quimio). The Chinese University Press, Hong Kong, pp. 363-382.

Bano, Z. Nagraja, N., Vibhakar, K.V. and Kapur, O.P. 1981. Minerals and heavy metal content in the cultivated mushrooms, *Agaricus bisporus*. Mush. News, **37**: 17-26.

Bano, Z., Ahmed, R. and Srivastava, H.C. 1964. Amino acids of edible mushrooms, *Lepiota* sp. and *Termitomyces* sp. Ind. J. Chem. **2**: 380-381.

Bano, Z., Bhagya, S. and Srinivasan, K.S. 1981. Essential amino acid compositions and proximate analysis of the mushrooms – *Pleurotus eous* and *Pleurotus florida*. Mushrooms News Letter for the Tropics. **1**: 6-10.

Bano, Z., Srinivasan, K.S. and Singh N.S. 1971. Essential amino acid composition of the proteins of mushroom (*Volvariella diplasia*). J. Food Sci. Techno. **8**: 180-182.

Bano, Z., Srinivasan, K.S. and Srivastava, H.C. 1963. Amino acid composition of the protein from mushrooms (*Pleurotus* sp.) Appl. Microbiol. **11**: 184-187.

Barnett, H.L. and Hunter, B.B. 1972. Illustrated genera of Imperfect Fungi. Burgess Publishing Company. pp. 241.

Baron, d'Yvoire. 1989. La morille, procede de culture applicable a tous les jardins. Rev. Sci. Nat. Appl.

Batra, L.R. and Batra, S.W.T. 1963. Indian Discomycetes. Univ. Kansas Sci. Bull. **44**: 109-256.

Bawa, A.S. and Khanum, F. 2003. Nutraceuticals: Mechanism of Action. Indian Food Indus. **22**: 44-51.

Beckett, A., Barton, R. and Wilson, I.M. 1968. Fine structure of wall and appendages formation in ascospores of *Podospora anserina*. J. Gen. Microbiol. **53**: 89-94.

Beelman, R.B. and Edwards, C.G. 1989. Variability in the composition and nutritional value of the cultivated mushroom *Agaricus bisporus*. Mush. News, **37**: 20-26.

Beelman, R.B. and Simons, S.S. 1996. Influence of selected cultural factors on relative tyrosinase activity in cultivated mushrooms. *In:* Mushroom Biology and Mushroom Product, *Proceeding of the 2nd International Conference* (Ed. D.J. Royse). The Pennsylvania State University, Pennsylvania, USA, pp. 543-551.

Beelman, R.B., Daniel, J.R. and Chikthimmah, N. 2003. Bioactive components in Button mushroom *Agaricus bisporus* (J. Lange) Imbach (Agaricomycetidae) of Nutritional, Medicinal, and Biological Importance. Intern. J. Med. Mush. **5**: 1-92.

Behl, N. 1987. Mushrooms in Pharmacology. Indian J. Mush. **12-13**: 42-49.

Belskonig, H.C., Gerrita, J.P.G. and Vaandrager, M.H. 1962. Some fungi appearing towards the end of composting. Mush. Sci. **5**: 165-169.

Berthet, P. 1964. Essais biotaxonomique sur les discomycetes. Ph.D. Thesis, Faculte des Sciences de 1' Universite de Lyon, France.

Bhandary, H.R. 1991. Some edible and medicinal fungi from Dumre to Manang, Mustang and Pokhara. J. Nat. Hist. Mush. **12**: 47-60.

Bhatt, R.P. 1986. Systematics and Ecobiology of some Agaric families. Ph.D. Thesis. Himachal Pradesh University, Shimla-171005.

Bhatt, R.P. and Lakhanpal, T.N. 1988a. *Russula crustosa* Peck.—An addition to Indian edible mushrooms. Curr. Sci. **57**:560-561.

Bhatt, R.P. and Lakhanpal, T.N. 1988b. *Amanita fulva* (Schaeff. ex. Pers.) - An edible mushroom new to India. Curr. Sci. **57**: 1126-27.

Bhattacharya, B. and Baruah, H.K. 1953. Fungi of Assam. J. Univ. Gauhati, **4**: 287-312.

Biesalski, H.K. 2001. Nutraceuticals: the link between nutrition and medicine. *In:* Nutraceuticals in health and disease prevention (Eds. K. Kramer, P.P. Hoppe and L. Packer). New York: Marcel Dekker Inc., pp. 1-26.

Binyami, N. 1980. Succession of Israel agaric flora. Nova Hedw. 32: 185-198.

Bisaria, R., Madan, M. and Bisaria, V.S. 1987. Mineral content of mushroom, *Pleurotus sajor-caju*, cultivated on different agrowastes. Mushroom J. Tropics, **7**: 53-60.

Blazs, S. and Szaboo, I. 1979. Temperature requirement studies on some new mushroom species in culture. Mush. Sci. **10**: 421- 427.

Block, S.S., Tsao, G. and Han, L. 958. Production of mushrooms from saw dust. J. Agr. Food Chem. **6**: 923-927.

Boa, E.R. 2002. How do local people make use of wild edible fungi? Personal narratives from Malawi. *In*: Edible ectomycorrhizal mushrooms and their cultivation (Eds. I.R. Hall, Y. Wang., A. Zambonelli and E. Daniell). Proceedings of the second international conference on edible ectomycorrhizal mushrooms. Christchurch, New Zealand, pp. 61.

Bobek, P., Ozdin, L. and Kuniak, L. 1993. Influence of water and ethanol extracts of the oyster mushroom *(Pleurotus ostreatus)* on serum and liver lipids of the Syrian hamster. Nahrung, **37**: 571-575.

Bok, J.W., Lermer, L., Chilton, J., Klingeman, H,G. and Towers, G.H. 1999. Antitumor sterols from the mycelia of *Cordyceps sinensis*. Phytochem. **51**: 891-898.

Borchers, A., Stern, J., Hackman, C. and Gershwin, M. 1999. Mushrooms, Tumors, and Immunity. Exp. Biol. Med. **221**: 281-293.

Boudier, E. 1897. Revision analytique des morilles de France. Bull. Soc. Mycol. Fr. **13**: 194-217.

Boudier, J.L.E. 1907. Histoire et classification des Discomycetes d' Europe. Paris.

Boyer, M. 1891. Note sur la reproduction des morelles. Bull. Soc. Mycol. France, **7**: 50.

Boyle, M. 2000. Module 1: Core Principles (Ed. Keith Hirst). P5. Harper Collins, London.

Bracker, C. 1969. Sporogenesis in Mucorales and Ascomycetes–XI. Intern. Bot. Congr. Abstracts, pp. 20.

Bresinky, A., Glasser, W. and Stangl, J. 1972. Untersuchungen zur Sippenstruktur der Morchellaceae. Berichte der Bayerischen Botanischen Gesellschaft, **42**: 127-143.

Bresinsky, A. and Stangal, J. 1961. Artabgrenzung innerhalb der Gattung *Morchella* 2. Pilzk. **27**: 102-110.

Brock, D. 1951. Studies on nutrition of *Morchella esculenta* Fries. Mycologia, **43**: 402-422.

Buller, A.H.R. 1914. The fungus lores of the Greeks and Romans. Trans. Br. Mycol. Soc. **5**: 21-66.

Bunyard, B.A., Nicholson, M.S. and Royse, D.J. 1994. A systematic assessment of *Morchella* using RFLP analysis of the 28S ribosomal RNA gene. Mycologia, **86**: 762-772.

Bunyard, B.A., Nicholson, M.S. and Royse, D.J. 1995. Phylogenetic Resolution of *Morchella, Verpa* and *Discoiotis* Boud. [Pezizales; Morchellaceae] based on restriction enzyme analysis of the 28s Ribosomal RNA gene. Exp. Mycol. **19**: 223-233.

Burnett, J.H. 1975. Mycogenetics. An Introduction to the general genetics of fungi. John Wiley & Sons, London.

Burnett, J.H. 1976. Fundamentals of Mycology. Edward Arnold Ltd. London, pp. 673.

Buscot, F. 1987. Contribution a l'edude d'une de morille de la foret rhename: *Morchella rotunda* (Pers.) Boudier Perspectives de domestication. These de Doctorat universite de Strasbourg, (France), pp. 179.

Buscot, F. 1989. Field observations on growth and development of *Morchella rotunda* and *Mitrophora semilibera* in relation to forest soil temperature. Can. J. Bot. **67**: 589-593.

Buscot, F. 1992. Mycorrhizal succession and morel biology. *In*: Mycorrhizas in Ecosystem (Eds. D.J. Royse, D.H. Lewis, A.H. Filler and J.J. Alexander). CAB International. pp. 220-224.

Buscot, F. 1993. Mycelial differentiation of *Morchella esculenta* in pure culture. Mycol. Res. **97**: 136-140.

Buscot, F. and Bernillon, J. 1991. Mycosporins and related compounds in field and cultured *Morchella esculenta*. Mycol. Res. **95**: 752-754.

Buscot, F. and Kottke, I. 1990. The association of *Morchella rotunda* with roots of *Picea abies*. New Phytologist, **116**: 425-430.

Buscot, F. and Roux, J. 1987. Association between living roots and ascocarps of *Morchella rotunda*. Trans. Br. Mycol. Soc. **89**: 249-252.

Buscot, F., Wipf, D., De Battista, C., Munch, J., Bolton, B. and Martin, F. 1996. DNA polymorphism in morels: PCR/RFLP analysis of the ribosomal DNA spacers and microsatellites primed PCR. Mycol. Res. **110**: 63-71.

Butler, E.J. and Bisby, G.R. 1931. The Fungi of India. Imp. Counc. Agri. Res. India, Sci. Mono. 1. Calcutta, pp. 237.

Butler, E.J. and Bisby, G.R. 1960. The fungi of India. Revised by R.S.Vasudeva, I.C.A.R., New Delhi, pp. 552.

Cailleux, R. 1956. Culture des Champignos sutres que le Champignos de couche. Mush. Sci. **3**: 272-283.

Cailleux, R. 1969. Pent on cultivar les morilles. Rev. de Mycol. **33**: 304-308.

Campbell, W.A. 1938. Preliminary report on decay in sprout northern hardwoods in relation to timber stand forest Exp. Sta. Occas. Paper 7, pp. 8.

Carroll, G.C. 1967. The ultrastructure of ascospore delimitation of *Saccobolus kerverni*. Arch. Microbiol. **33**: 218-224.

Carroll, G.C. 1969. A study of fine structure of ascosporogenesis in *Saccobolus kerverni*. Arch. Microbiol. **66**: 321-339.

Carroll, G.C. and Carroll, F.E. 1974. The fine structure of conidium development in *Phialophora dimorphospora*. Can. J. Bot. **52**: 2119-2128.

Cassileth, B.R. 2000. Complementary Therapies: The American experience. Supportive Care, Cancer, **8**:16-23.

Cavazzoni, V. and Manzoni, M. 1994. Enzyme analysis. Lebensmittel-Wissenschaft and Technologie, **27**: 73-77.

Chadha, K.L. 1992. Mushroom Research and Developments in India. Mush. Res. **1**: 1-12.

Chadha, K.L. and Sharma, S.R. 1995. Mushroom Research in India – History Infrastructure and Achievements. *In*: Advances in Horticulture. Vol. 13 – Mushroom (Eds. K.L. Chadha and S.R. Sharma) Malhotra Publishing House, New Delhi. pp. 1-33.

Champion, H.C. and Seth, P.K. 1968. A revised Survey of the forest types of India, F.R.I. Dehradun, Government of India, Press, Nasik, pp. 404.

Chang, A.W. and Miles, P.G. 1996. Biomedical Research and the Application of mushroom nutriceuticals from *Ganoderma lucidum*. *In:* Mushroom biology and Mushroom Products (Ed. D.J. Royse). The Pennsylvania State University, Pennsylvania, USA, pp. 161-175.

Chang, S.T. 1972. The Chinese Mushrooms. The Chinese University Press, Hongkong, pp. 41.

Chang, S.T. 1980. Mushroom as Human Food. Bioscience, **30**: 399-401.

Chang, S.T. 1982. The prospects for mushroom protein in developing countries. *In*: The Tropical Mushrooms: Biological Nature and Cultivation Methods (Eds. S.T. Chang and T.H. Quimio). The Chinese University Press, Hong Kong, pp. 463-473.

Chang, S.T. 1993. Mushroom Biology: The impact on mushroom production and mushroom products. *In:* Mushroom Biology and Mushroom Products (Eds. S.T. Chang, J.A. Buswell and S.W. Chiu). The Chinese University Press, Hong Kong, pp. 3-20.

Chang, S.T. 1999. Global impact of edible and medicinal mushrooms on human welfare in 21st century: Nongreen Revolution. Intern. J. Med. Mush. **1**: 1-8.

Chang, S.T. and Buswell J.A. 1996. Mushroom Nutriceuticals. *World* J. Microbiol. Biotechnol. **12**: 473-476.

Chang, S.T. and Buswell, J.A. 1999. *Ganoderma lucidum* (Curt.: Fr.) P. Karst. (Aphyllophoromycetideae) - A mushrooming medicinal mushroom. Intern. J. Med. Mush. **1**: 139-148.

Chang, S.T. and Hayes, W.A. 1978. The Biology and Cultivation of edible Mushrooms. Acad. Press, New York, pp. 819.

Chang, S.T. and Miles, P.G. 1982. Introduction to Mushroom Science. *In*: The Tropical Mushrooms: Biological Nature and Cultivation Method (Eds. S.T. Chang and T.H. Quimio). The Chinese University Press, Hong Kong, pp. 3-11.

Chang, S.T. and Miles, P.G. 1989. Edible mushrooms and their cultivation. CRC Press. Boca Raton, pp. 345.

Chang-Ho, Y. and Yee, N.T. 1977. Comparative study of the physiology of *Volvariella volvacea* and *Coprinus cinereus*. Trans. Br. Mycol. Soc. **68**: 167-172.

Charaya, M.U. and Mehrotra, R.S. 1999. From Ethnomycology to Fungal Biotechnology: A Historical Perspective. *In*: From Ethnomycology to Fungal Biotechnology: Exploiting Fungi from Natural Resources for Novel Products (Eds. J. Singh and K.R. Aneja). Kluwer Academic Publishers, New York, pp. 1-10.

Chaturvedi, A. 1987. Cultural and Cultivation studies on *Pleurotus* species from Himachal Pradesh. M.Phil. Dissertation, Himachal Pradesh University, Shimla-171005.

Chauhan, A. 2003. Studies on some wild edible mushrooms of Distt. Kinnaur. M.Phil. Dissertation, Himachal Pradesh University, Shimla–171 005.

Chen, Q. 1989. Antilipemic effects of polysaccharides from *Auricularia auricula, Tremella fuciformis,* and *T. fuciformis* spores. Zhongguo Yaoke Daxue Xuebae, **20**: 344-347.

Chen, S.Z. and Chu, J.Z. 1996. NMR and IR studies on the characterization of Cordycepin and 2'deoxyadenosine. Zhongguo Kangshensu Zaxhi. **21**: 9-12.

Cheung, P.C.K. 1996. Dietary fibre content and composition of some cultivated edible mushrooms fruiting bodies and mycelia. J. Agri. Food Chem. **44**: 468-471.

Cheung, P.C.K. 1997. Dietary fibre content and composition of some edible fungi determined by two methods of analysis. J. Sci. food Agri. **73**: 255-260.

Chihara, G. 1992. Immunopharmacology of Lentinan, a polysaccharide isolated from *Lentinus edodes*: Its application as a host defence potentiator. Inter. J. Oriental Med. **17**: 57-77.

Chihara, G. 1993. Medical aspects of Lentinan isolated from *Lentinus edodes* (Berk.) Sing. *In*: Mushroom Biology and Mushroom Products (Eds. S.T. Chang, J.A. Buswell and S.W. Ehiu). The Chinese University Press, Hong Kong, pp. 221-226.

Chihara, G., Maeda, Y.Y., Hamuro, J., Sasaki, T. and Fukuoka, F. 1969. Inhibition of mouse Sarcoma 180 by polysaccharides from *Lentinus edodes* (Berk.) Sing. Nature, **222**: 687-688.

Cho, S.M., Lee, J.H., Han, S.B. and Kim, B.K. 1999. Chemical features and purification of immunostimulating polysaccharides from the fruiting bodies of *Agaricus blazei* Murill. Korean J. Mycol. **27**: 170-174.

Choudhary, M.S., Ahmad, A., Ali, Sher, H. and Malik, S. 2000. Technical Report on Market Study of Medicinal Herbs in Malakhand, Peshawar, Lahore and Karachi. SDC - Interco Operation, Peshawar. Retrieved from the website: http/www.ethnobotanical reports/medicinal herbs/ morels/html.

Christensen, C.M. 1966. The Molds and Man. University of Minnesota, Minneapolis, pp. 65.

Clark, L.C., Combs, G.F., Turnbull, B.W., Slate, E.H., Chalker, D.K., Chow, J., Davis, L.S., Glover, R.A., Graham, G.F., Gross, E.G., Krongrad, A., Lesher, J.L., Park, H.K., Sanders, B.B., Smithm, C.L. and Taylor, J.R. 1996. Effects of selenium supplementation for cancer prevention on patients with carcinoma of the skin. J. Am. Med. Assoc. **276**: 1957-1963.

Claussen, P. 1912. Zur Entwicklugsge - Schichteder Ascomyceten *Pyronema confluens*. Zeitschr Botanik. **4**: 1-64.

Cochrane, V.W. 1958. Physiology of Fungi. John Wiley and Sons, Inc., New York, pp. 524.

Cooke, M.C. 1870. Kashmir Morels. Trans. Bot. Soc. Edinburgh, **10**: 439-443.

Cooke, M.C. 1876. Some Indian Fungi. Grev. **4**: 114-118.

Coppens, P. 2002. Food of the Gods? Retrieved from the website: http://www.philipcoppens.com/ mushrooms.html.

Corner, E.J.H. 1934. the Fungi of Wicken Fen. Cambridgeshire. Trans. Br. Mycol. Soc. **19**: 280.

Costantin, J. 1936. La culture la morille et sa forme condienne. Ann. Sci. Nat. Bot. **18**: 111-140.

Cotter, H.V.T. 1987. The systematics and Ecology of Boletes with special reference to the genus *Suillus* and its ectomycorrhizal relationships in Nepal. Ph.D. Thesis, Virginia Polytechnic Institute and State University, U.S.A. pp. 176.

Crisan, E.V. and Sands, A. 1978. Nutritional Value. *In*: The Biology and Cultivation of Edible Mushrooms (Eds. S.T. Chang and W.A. Hayes), Academic Press, London, pp. 137-165.

Das, L. 2003. Cooking God's own food - the South Indian way. *In*: Current Vistas in Mushroom Biology and Production (Eds. R.C. Upadhyay, S.K. Singh and R.D. Rai). Mushroom Society of India, pp. 253-256.

Dasvila, E.J, Buswell, J.A. and Chang, S.T. 2003. Bioceuticals - Health Aids. Agro. Food Indus. Hi-Tech. pp. 21-25.

Davey, C.B. 1969. Non-pathogenic organisms associated with mycorrhizae. *In*: Proc. First N. American Conf on Mycorrhizae. U.S.D.A. Misc. Publ. 1189, pp. 114-121.

Dean, J.A. 1960. Flame Photometry. McGraw-Hill Book Co. Inc.

Delmanto, R.D., de Lima P.L., and Hutchinson, R.M. 2001. Antimutagenic effect of *Agaricus blazei* Murill. mushroom on genotoxicity induced by cyclophosphamide. Mutat. Res. **496**: 15-21.

Delmas, J. 1974. Less morilles, caracteres, ecologie researchers en cours. Pepinieres, Horticulture et Maraichage, **149**: 23-30.

Delmas, J. 1978. The potential cultivation of various edible fungi. *In*: The Biology and Cultivation of Edible Mushrooms (Eds. S.T. Chang and W.A. Hayes). Academic Press, London, pp. 699-722.

Delmas, J. and Bunel, F. 1974. Contribution a' L' etude ecologigne des morilles. C.R. Seances A. Lad. Agri. Fr. **23**: 118-125.

Delmas, J. and Poitou, N. 1974. Introduction a L'ecologie des morielles en France. Mush. Sci. **9**: 847-857.

Dennis, R.W.G. 1960. British cup – Fungi and their Allies. The Ray Society, London, pp. 280.

Dennis, R.W.G. 1968. British Ascomycetes, J. Cramer Lehre, Germany.

Dermek, A. 1982. Mushroom and other fungi. Slovart Publishers, England, pp. 25-39.

Dhar, B.L. and Munjal, R.L. 1978 Microflora of Synthetic compost for *agaricus bisporus*. Ind. J. Mush. **1**: 87-92.

Dibben, M. 1996. Mushrooms Mania.........is it for you? *In*: The Field Museum of Natural History Bulletin, Field Museum, Chicago, Illinois. Retrieved from the website: http/www.dibben mushroom mania.com/mushrooms.html.

Dillenius, J.J. 1719. Catalogues plantrum sponte cira *Gissam nascentium*. Cum appendice, Frencofurtied Moenum.

Dizbay, M. and Karaboz, I, 1986. Submerged Culture and Ascocarp Production of *Morchella* spp. Doya Biyol Serisi, **10**: 326-330.

Doyle, J.J. and Doyle, J.J. 1987. A rapid DNA isolation procedure for small quantities of fresh leaf tissue. Phytochem. Bulletin, **19**: 11-15.

Dubois, M., Glles, K.A., Hamilton, J.K., Roberts, P.A. and Smith, F. 1956. Calorimetric method for the determination of sugars and related substances. Determination of Concentration of pure Sugar Solution. Anal. Chem. **28**: 350-360.

Dudka, I.A. 2001. Mushrooms in folk medicine of the Eastern Slavs. Intern. J. Med. Mush. **3**: 135.

Dyke, A.J. and Newton, A.C. 1999. Commercial harvesting of wild mushrooms in Scottish forests: is it sustainable? Scottish Forestry, **53**: 77-85.

Dymock, W., Warden, C.H.J. and Hooper, D. 1890. Pharmacographia Indica – A History of the Principal Drugs of Vegetable Origin, Part III, Education Society Press, Byeulla, Bombay, pp. 629-635.

Eckblad, F.E. 1968. The genera of Operculate Discomycetes. *In*: A re-evaluation of their Taxonomy, Phylogeny and Nomenclature. Nylt. Mag. Bot. pp. 1-192.

Falck, R. 1920. Wage Zur Kultur der Morchelarten-Pilze-und Krauter. Fereund, **3**: 211-223, 247-255.

Faulks, P.J. 1958. An introduction of Ethnobotany. Moredale Publication, London.

Fergus, C.L. 1964. Thermophilic and Thermo-tolerant moulds and actinomycetes of Mushroom compost during peak heating. Mycologia **56**: 267-284.

Filipov, D. 1998. Mushroom season has Russians in fungi frenzy. Boston Globe. Retrieved from the website: http/www.mushroom season in fungi.com/mushrooms.html.

Fincham, J.R.S. 1949. Chromosome number in species of *Neurospora*. Ann. Bot. (London). **13**: 23-26.

Fincham, J.R.S., Day, P.R. and Radford, A. 1979. Fungal Genetics. Univ. California Press, Berkely, pp. 312-318.

Findlay, W.P.K. 1982. Fungi, Folklore, Fiction and Fact. The Richmond Publishing Company Ltd., U.K.

Fitzpatrick, W.H., Esselen, W.B. Jr., and Weir, E. 1946. Composition and nutritive value of Mushroom protein. J. Am. Diet. Assoc. **22**: 318-323.

Flores, R., Bran, M. D.C. and Honrubia, M. 2002. Edible mycorrhizal mushrooms of the west Highland Guatemala. *In:* Edible ectomycorrhizal mushrooms and their cultivation (Eds. I.R. Hall, Y. Wang, A. Zambonelli and E. Daniell). Proc. Second international conference on edible mycorrhizal mushrooms. Christchurch, New Zealand Institute.

Fogarty, W.M. and Kelly, C.T. 1979. Starch degrading enzymes of microbial origin. Indus. Microbiol. **15**: 87-105.

Food and Agricultural Organisation, 2004. Wild edible fungi: A global overview of their use and importance to people, Rome, Italy. Retrieved from the website: http://www.fao.org/docrep/007/y5489e/y5489e00.htm.

Food and Agricultural Organization. 1970. Amino acid Content of Foods and Biological Data on Proteins. Nutr. Stud. No. 24. Food Policy and Food Sci. Serv., Nutr. Div., Food Agric. Organ. U.N. Rome.

Food and Agricultural Organization. 1972. Food Composition table for use in East Asia (Part - 1). Food Policy and Nutrition on Div., Food Agaric Organ. U.N. Rome.

Food and Agricultural Organization. 1973. Energy and Protein Requirements. Report of a joint FAO/WHO ad hoc.Expert Committee Food. Nutr. Meet. Rep. Ser. No. 52. Food Agric. Organ. U.N. Rome.

Foster, R.C. 1986. The ultrastructure of rhizoplane and rhizosphere. Ann. Rev. Phytopathol. **24**: 211-234.

Foster, R.C., Marks, G.C. 1967. Observations on the Mycorrhizae of Forest Trees. II. The rhizosphere of *Pinus radiata*. Don., Aust. J. Biol. Sci., **20**: 915-926.

Frank, A.B. 1885. Uber die auf/Wurzelymbiose beruhende. Ernahrung gewisser Baume durch Uterirdische Pilze Berdt. Bot. Ges. **3**: 128-135.

Fries, E.M. 1822. "Systema Mycologicum" Vol. 3. Berlingiana, Lund, Sweden, pp. 238.

Fries, E.M. 1829. Systema Mycologicum. Vol. 3. Gryrhiswaldial, pp. 1-524.

Fron, G. 1905. Sur les conditions de development du mycelium de la morille. Compt. Rend. Acad. Sci. Paris, **140**: 1187-1189.

Fujimiya, Y., Suzuki, Y., Oshiman, K., Kobori, K., Morigushi, K., Nakashima, H., Matumoto, Y., Takahara, S., Ebina, T. and Katakura, R. 1998. Selective tumoricidal effect of soluble proteoglucan extracted from the basidiomycete, *Agaricus blazei* Murill. mediated via natural killer cell activation and apoptosis. Cancer Immunol. Immunother. **46**: 147-159.

Fujimiya, Y., Yamamoto, H., Niji, M. Suzuki, I. 2000. Peroral effect on tumor progression of soluble beta (1,6) glucans prepared by acid treatment from *Agaricus blazei* Murrll. (Agaricaceae, Higher Basidiomycetes). Intern. J. Med. Mush. **2**: 43-50.

Gao, Y.H. 2000. The miracle herb, scientific reports of *Ganoderma*. Yuangizai Publisher, Taipei.

Garrett, S.D. 1960. Biology of Root Infecting Fungi Univ. of Cambridge Press, Cambridge, pp. 293.

Gerdmann, J.W. 1968. Vesicular-Arbuscular Mycorrhiza and plant growth. Ann. Rev. Phytopathol. **6**: 397-418.

Gessner, R.V., Romano, M.A. and Schultz, R.W. 1987. Allelic variation and segregation of *Morchella deliciosa* and *Morchella esculenta*. Mycologia, **79**: 683-687.

Ghosh, R.N. and Pathak, N.C. 1962. Fungi of India-1 *Morchella, Verpa,* and *Helvella.* Bull. Nat. Bot. Gardens (Lucknow), **71**: 1-19.

Ghurde, V.R. and Wakode, D.D. 1981. A new report of *Morchella* from Central India. Indian J. Mycol. Plant Pathol. **11**: 314-315.

Gilbert, F.A. 1960. The submerged culture of *Morchella.* Mycologia, **52**: 201-205.

Gilbert, F.A. and Robinson, R.F. 1957. Food from Fungi. Econ. Bot. **11**: 126-145.

Gill, F. 1973. Mesosomes: their role in the delimitation of the ascospores. Mycopathol. Mycol. Appl. **49**: 243-247.

Gilman, J.C. 1967. Manual of soil Fungi. Oxford and IBH Publishing, Co. pp. 450.

Gordon, D.T. 2002. Intestinal health through dietary fibre, prebiotics, and probiotics. Food Tech. **56**: 23.

Gowry, M. 1990. Studies on some general aspects of genus *Morchella.* M.Phil. Dissertation, Himachal Pradesh University, Shimla-171005.

Graham, V.O. 1927. Ecology of the fungi in the Chicago region. Bot. Gaz. 267-287.

Grainger, T. 1946. Ecology of the Larger Fungi. Trans. Br. Mycol. Soc. **29**: 52-63.

Gray, W.D. 1959. The Relation of Fungi to Human Affairs. Hennery Holt and Co., New York, pp. 510.

Gray, W.D. 1970. The Use of Fungi as Food and in Food Processing. Butterworth, London, pp. 113.

Greenhalgh, G.N. and Griffiths, H.B. 1970. The ascus vesicle. Trans. Br. Mycol. Soc. **54**: 489-492.

Greish, H. 1940. Befruchtungsarten bei *Morchella* fur wissenschaftliche. Botanik, **89**: 245-253.

Gries, H. 1940. Befruchtungsarten bei *Morchella.* Jahrb. Wiss. Bot. 89: 245-253.

Groves, J.N. and Hoare, S.C. 1953. The Helvellaceae of Ottawa district, Canada. Fld. Nat. **67**:95-102.

Gunde-Cimerman, N. 1999. Medicinal value of the genus *Pleurotus* (Fr.) P. Karst. (Agaricales S.R., Basidiomycetes). Intern. J. Med. Mush. **1**: 69-80.

Gupta, S.P. 1985. Statistical Methods. Sultan Chand and Sons, New Delhi.

Gupta, Y. 1990. *Morchella*: Recent Advances in Cultivation. *In*: Recent developments in the cultivation technology of edible mushrooms (Eds. S.R. Sharma, B.L. Dhar and K.B. Mehta). National Centre for Mushroom Research and Training,, Chambaghat, Solan, pp. 252-256.

Guzman, G. 2001. Medicinal fungi in Mexico: Traditions, Myths and Knowledge. Intern. J. Med. Mush. **3**: 95.

Halpern, G. 1999. *Cordyceps*, China's healing mushroom. Avery Publishing, **1999**: 63-70.

Hamayun, M., and Khan, M.A. 2004. Studies on collection and marketing of Morchella (morels) Uttor-Gabral Velleys, Swat, Pakistan. Retrieved from the website: http://www.siu.edu/~ebl/leaf lets/ morels.htm.

Hamayun, M., Khan, M.A. and Begum, S. 2003. Marketing of medicinal plants of Uttor-Gabral Velleys, Swat, Pakistan. J. Ethnobot. Leaflets, SIUC, USA, pp. 1-20.

Harkonen, M. 1998. Uses of mushrooms by Finns and Karelians. Intern. J. Circumpolar Health, **40**: 40-55.

Harkonen, M. 2002. Mushroom collecting in Tanzania and Hunan (southern China): inherited wisdom and folklore of two different cultures. *In*: Tropical Mycology, Vol. 1 Macromycetes (Eds. R. Watling, J.C. Frankland, A.M. Ainsworth, S. Isaac and C.H. Robinson). Wallingford, UK, CAB International, pp. 149-165.

Harkonen, M., Saarimaki, T. and Mwasumbi, L. 1994a. Edible and poisonous mushrooms of Tanzania. The African J. Mycol. Biotechnol. **2**: 99-123.

Harkonen, M., Saarimaki, T. and Mwasumbi, L. 1994b. Tanzanian mushrooms and their uses. Some reddish edible and poisonous *Amanita* species. Karstenia, **34**: 47-60.

Harley, J.L. 1948. Mycorrhiza and Soil Ecology. Biol. Rev. **23**: 127-158.

Harper, R.A. 1897. Kernteilung and freice Zell-bildung in Ascus. Jahrb. Wiss. Botan. **30**: 249-284.

Harper, R.A. 1907. Sexual reproduction and the Organization of the nucleus in certain mildews. Carngia Inst. Wash. Publ. **37**: 104.

Harshberger, J.W. 1895. Some new ideas: The plants cultivated by aboriginal people and used in primitive commerce. The Evening Telegraph (daily), Philadelphia, **63**: 2.

Harshberger, J.W. 1896. The Purpose of Ethnobotany. Bot. Gazette, **34**: 146-154.

Hatanaka, S.I. 1969. A new amino acid isolated from *Morchella esculenta* and related species. Phytochem. **8**: 1305-1308.

Hathcock, J. 2001. Dietary Supplements: How they are used and regulated? J. Nutrition, **131**: 1114-1117.

Hatvani, N. 2001. Antibacterial effects of the culture fluid of *Lentinus edodes* mycelium grown in submerged liquid culture. Intern. J. Antibacterial Agents, **17**: 71-74.

Hawksworth, D.L., Kirk, P.M., Sutton, B.C. and Pegler, D.N. 1995. Ainsworth & Bisby Dictionary of Fungi (8th ed. International Mycological Institute), CAB International.

Hayes, W.A. and Haddad, N. 1976. The food value of cultivated mushrooms and its importance to the mushroom industry. Mush. J. **40**: 104-106.

Heim, R. 1936. La culture des morilles. Rev. Mycol. Paris, Suppl. **1**: 10-11, **2**: 19-25.

Heim, R. 1978. Les Chamignons Toxiques et Hallucinogens. Boubee, Paris.

Heim, R., Cailleux, R., Wasson, R.G. and Thevenard, P. 1967. Nouvelle investigations sur les Champignons Hallucinogenes, Paris (Mus. Nat. Hist. Natur.).

Heiser, C.B. 1995. Edgar Anderson botanist and curator of useful plants. Ann. Miss. Bot. Gard. **82**: 54-60.

Hennings, P. 1901. Fungi Indiae Orientalis II, C.W. Gollana. 1900. Collecti. Hedw. **40**: 323-42.

Hervey, A., Bislig, G. and Leong, J. 1978. Cultural studies of single ascospore isolates of *Morchella esculenta*. Mycologia, **70**: 1269-1274.

Hiroshi, S. and Takeda, M. 1993. Diverse biological activity of PSK (Krestin), a protein-bound polysaccharide from *Coriolus versicolor* (Fr.) Quel. *In:* Mushroom Biology and Mushroom Products (Eds. S.T. Chang, J.A. Buswell and S.W. Chiu). The Chinese University Press, Hong Kong.

Hobbs, C. 1986. Medicinal Mushrooms: An exploration of traditional, healing and culture. Botanica Pres, Santa Cruz, CA, pp. 81-82.

Hobbs, C. H. 1995. Medicinal Mushrooms: An exploration of traditional, healing and culture (2nd ed.). Botanica Press, CA. pp. 172.

Hokama, Y. and Hokama, J.L. 1981. *In vitro* inhibition of platelet aggregation with low dalton compounds from aqueous dialysates of edible fungi. Research communications in Chemistry, Pathology and Pharmacology, **31**: 17-180.

Huang, B.H., Yung, K.H. and Chang, S.T. 1985. The sterol composition of *Volvariella volvaceae* and other edible mushrooms. Mycologia, 77: 954.

Hung, C.G. and Wells, K. 1917. Light and electron microscopic studies of crozier development in *Pyronema domesticum.* J. Gen. Microbiol. **66**: 15-27.

Hutchinson, L.J. and Malloch, D.W. 1988. A verification of protocol for cultural isolates of ectomycorrhizal basidiomycetes.

Ikewawa, J., Uehara, N., Maeda, Y., Nakanishi, M. and Fukuoka, F. 1969. Antitumor activity of aqueous extract of some edible mushrooms. Cancer Res. **29**: 734-735.

Imai, S. 1932. Contribution to knowledge of the classification of Helvellaceae. Bot. Mag. Tokyo, **48**: 172-177.

Imai, S. 1954. Elvellaceae Japoniae. Sci. Rep. Yokohama. Natn. Univ. Ii. **3**: 1-35.

Iqbal, M. 1993. International trade in non-wood forest products. An Overview. FAO, Rome.

Ito, H., Shimura, K. and Kawade, M. 1997. Antitumor effects of a new polysaccharide - protein complex (ATOM) prepared from *Agaricus blazei* Murill. (Iwade Starin 101) "Himematsutake" and its mechanism in tumours bearing mice. Anticancer Res. **17**: 277-284.

Jackson, M.L. 1958. Soil Chemical Analysis. Prentice-Hall, Inc. pp. 498.

Jackson, M.L. 1967. Soil Chemical Analysis. Asia Publishing House, Bombay, pp. 498.

Jackson, R.M. 1965. Antibiotics and Fungistatics of soil microorganisms. *In*: Ecology of soil borne plant pathogens: Biological control (Eds. K.F. Baker and W.C. Synder), Univ. of Calif. Press, Berkeley, pp. 363-373.

Jacobson J.S., Workman, S.B., and Korenberg, F. 2000. Research on complementary/alternative medicine for patients with breast cancer: A review of the biomedical literature. J. Clinical Oncology **18**: 668-683.

Jacobson, E. 2003. Does winter snow make mushrooms grow? Retrieved from the website: www. mushroom expert.com/msonsoonmorels.htm.

Jacquetant, E. 1984. Les Morilles. La Biblitheque des Arts, Paris, pp. 114.

Jain, A.K. 2004. Studies on Mushroom Cultivation with special reference to *Pleurotus* species and their marketing potential in Sagar region. Ph.D. Thesis, Dr. Hari Singh Gaur University, Sagar, M.P. 470 003.

Jain, S.K. 1981. Glimpses of Ethnobotany. Oxford and IBH, New Delhi, pp. 365.

Jain, S.K. 1995. Ethnobotanical diversity in Zingibers of India. Ethnobotany, **1**: 83-88.

Janardhan, K.K. and Hussain, A. 1978. Nutritive value of two *Morchella* strains grown in a cauliflower leaf extract cornsteep medium. Ind. J. Microbiol. **15**: 84.

Jandaik, C.L. and Kapoor, J.N. 1975. Nutritive value of the mushroom *Pleurotus sajor-caju*. Mush. J. **36**: 408-410.

Jandaik, C.L. and Sharma, S.R. 1995. Present status of *Morchella* in India. *In*: Advances in Horticulture (Eds. K.L. Chadda and S.R. Sharma). Malhotra Publishing House, New Delhi, pp. 171-194.

Jandaik, C.L., Bhandari, A. and Arora, C.L. 1978. Chemical composition of some edible fungi. Mush. Sci. **10**: 684-688.

Jandaik, C.L. Bhandari, A.R. Arora, C.L. and Ranged, C.O. 1979. Chemical composition of some edible fungi. Mush. Sci. **10**: 685-688.

Joshi, A.R. 1988. A Study of the Environmental Relationship of Certain Village Communities in the Central Development Region of Nepal. Ph.D. Thesis, University of Wales, U.K.

Joshi, A.R. 1991. A Study of Phyto-ecology of Sundarijal and Adjoining Areas (Shivapuri Watershed and Wildlife Reserve), Snowy Mountain. Engineering Corp. Ltd.

Joshi, A.R. 1992. Environmental Dimension of ESSD: A Case Study of Hill Ecosystem. ESCAP, Bangkok, Thailand.

Joshi, A.R. 1993. EIA of Kali gandaki Hydropower Project (Biodiversity Report), East Consult, Kathmandu, Nepal.

Joshi, K. 1996. Uses of wild plants by local communities of Kakani area. Network (A newsletter of APINMAP/SCAMAP), Nepal, **1**: 2-3.

Joshi, K.L. and Joshi, A.R. 1999. Ethnobotanical study of some wild mushrooms of two valleys (Kathmandu and Pokhara) of Nepal. Ethnobotany, **11**: 47-56.

Joshi, K.L. 1984. Geography of Himachal Pradesh. National Book Trust of India, pp. 243.

Joshi. A.R., Adhikari, M.K., Joshi, K. and Adhikari, K. 1996. A study of food value and toxins of wild mushrooms of Kathmandu and Pokhara Valley. A News letter of APINMAP/SCAMAP, Nepal.

Jung, S.W., Gessner, R.V., Kekudell, K.C. and Romano, M.A. 1993. Systematics of *Morchella esculenta* complex using Enzyme-Linked Immunosorbent Assay. Mycologia, **85**: 677-684.

Kaisth, K. 1985. Studies on edible species of *Lactarius* from Himachal Pradesh. M.Phil. Dissertation, Himachal Pradesh University, Shimla - 171005.

Kalberer, P. and Kunsch, U. 1974. Amino acid composition of the oyster mushroom (*Pleurotus ostreatus*). Lebensm - Wiss. Technol. **7**: 242-244.

Kamal, S., Singh, S.K. and Tiwari, M. 2004. Role of enzymes in initiating sexual cycle in different species of *Morchella*. Indian Phytopath. **57**: 18-23.

Kanayian, S. and Ramaswamy, K. 1980. A Hand Book of Edible Mushrooms. Today and Tommorow Printers and Publishers, New Delhi, pp. 104.

Kaul, T.N. 1975. Studies on genus *Morchella* in Jammu & Kashmir-1. Soil Composition in relation to carpophore development. Bull. Soc. Bengal, **29**: 127-134.

Kaul, T.N. 1976. Common edible mushrooms of Jammu and Kashmir. Mush. Sci. **1**: 517-529.

Kaul, T.N. 1977a. Physiological studies on *Morchella* species I. Carbon utilization. Bull. Bot. Soc. Bengal, **31**: 35-42.

Kaul, T.N. 1977b. Physiological studies on *Morchella* species II. Nitrogen utilization. Indian J. Mush. **58**: 2-6.

Kaul, T.N. 1978. Nutritive value of some edible Morchellaceae. Ind. J. Mush. **4**: 26-34.

Kaul, T.N. 1981. Common edible mushrooms of Jammu and Kashmir. Indian Mush. Sci. **11**: 79-82.

Kaul, T.N., Kachroo, J. L. and Raina, A. 1976. Common edible mushrooms of Jammu and Kashmir. *In*: Proceedings of the First symposium on Survey and Cultivation on Edible Mushrooms in India. Reg. Res. Lab. Srinagar, **2**: 135-150.

Kaul, T.N., Khurana, M.L., Kachroo, J.L., Krishna, A. and Atal, C.K. 1981. Myco-ecological studies on morel bearing sites in Kasmir. Mush. Sci. **11**: 789-797.

Kaushal, S.C. 1986. *Morchella* – A fascinating mushroom. *In*: Souvenir on Mushrooms. National Centre for Mushroom Research and Training, Chambaghat, Solan, pp. 85-88.

Kaviyasaran, V., Kumar, M., Shiva, R., and Natrajan, K. 2006. *Morchella esculenta*: A new record from South India. Mushroom Research. **15**: 87-88.

Kawagishi, H., 1995. Isolation of 5′-deoxy-5′methyl sulfinyladenosine as platelet aggregation inhibitor from *Ganoderma lucidum*. *In*: Recent Advances in *Ganoderma lucidum* research (Eds. B.K. Kim, I.H. Kim and Y.S. Kim) Pharmaceutical Society of Korea, Seoul, Korea, pp. 61-92.

Kawagishi, H., Ando, M. and Mizuno, T. 1990. Hericenone A and B as cytotoxic principles from the mushroom *Hericium erinaceum*. Tetrahedron Letters, **31**: 373-376.

Khader, V. and Padmavati. 1991. Studies on Oyster mushroom (*Pleurotus* spp.) at different stages of maturity and standardization of three recipes from matured mushrooms. *In*: Indian Mushrooms – 1991 (Ed. M.C. Nair). Proceedings of National Symposium on Mushrooms. Kerala Agricultural University, Kerala, pp. 221-223.

Khanna, P.K. and Garcha, H.S. 1981. Nutritive value of mushroom *Pleurotus florida*. Mush. Sc. **11**: 561-572.

Kim, H.W. and Kim, B. K. 1999. Biomedicinal triterpenoides of *Ganoderma lucidum* (Curt.: Fr.) P. Krast. (Aphyllophoromycetideae). Intern. J. Med. Mush. **1**: 121-138.

Kimbrough, J.W. 1970. Current trends in the classification of Discomycetes. Bot. Rev. **36**: 91-161.

Kimura, Y., Tsuda, K., Arichi, S. and Takahashi, N, 1994. Clinical evaluation of sizofiran as immunotherapy in treatment of head and neck cancer. Acta. Otolargynzology, **511**: 192-195.

King, H.J. and Jacobson, K.M. 2000. Genetic diversity within and between populations of morel mushrooms (*Morchella esculenta*). Iowa Academy of Science 2000. Annual meeting, Des Mories.

Kino, K.Y., Yamaoka, K., Watanabe, J., Shimizu, K. and Tsunoo, H., 1989. Isolation and characterization of a new immunomodulatory protein Ling Zhi-8 (LZ-8) from *Ganoderma lucidum*. J. Biol. Chem. **264**: 472-478.

Kino, T., Tabata, H., Ukai, S. and Hara, C. 1986. A minor protein-containing galactomannan from a sodium carbonate extracts of *Cordyceps sinensis*. Carbohy. Res. **156**: 189-197.

Korf, R.P. 1972. Synoptic key of the genera of the Pezizales. Mycologia, **64**: 937-993.

Korf, R.P. 1973. Discomycetes and Tuberales. *In*: The Fungi, Vol. IV A (Eds. G.C. Ainsworth, F.K. Sparrow and A.S. Sussman). Acad. Press, New York, pp. 249-319.

Kornnerup, A. and Wanscher, J.H. 1978. Methuen Handbook of colour. Eyre Methuen, London, pp. 252.

Krieger, L.C.C. 1967. The Mushroom Handbook. Dover Publications. New York.

Kubo, K. and Nanba, H. 1996. Anti-diabetic mechanism of Maitake (*Grifola frondosa*). *In*: Mushroom Biology and Mushroom Products (Ed. D.J. Royse) Pennsylvania State University, Press, pp. 215-221.

Kulkarni, R.K. and Kamerath, C.D. 1989. Isozyme analysis of *Morchella* species. Mush. Sci. **12**: 451-457.

Kumar, A. 1987. Studies on some mushroom families in the N.W. Himalayas. Ph.D. Thesis. Himachal Pradesh University, Shimla-171005.

Kumar, A., Lakhanpal, T.N. and Stephenson, S.L. 1990. Ecological studies on some macro fungi in the north-western Himalayas. Nova Hedw. **50**: 535-547.

Kuo, M. 2002a. Misconceptions on picking morels. Retrieved from the website: http//www.bluewillowpages.com/mushroomexpert/morels/picking.html.

Kuo, M. 2002b. Preserving and drying morels. Retrieved from website: http//www.bluewillowpages.com/mushroomExpert/morels/drying.html

Kuo, M. 2002c. When and where morels grow. Retrieved from website: http//www.bluewillowpages.com/mushroomexpert/morels/when and where.html.

Kuo, M. 2003. West coast morels. Retrieved from the website: http//www.bluewillowpages.com/mushrooms/morels/west_coast.html.

Kurtzman, R.H. 1975. Mushrooms as a source of food protein. *In*: Protein Nutritional Quality of Foods and Feeds. Part II (Eds. M. Mriedman and Marcel Decker) Inc., New York, pp. 315.

Kurtzman, R.H. Jr. 1976. The metabolism of fatty acid substances by the oyster mushroom. Mush. Sci. **9**: 557-565.

Lachance, P.A. 2003. The Nutraceutical Institute. Retrieved from the website: file://A:/a/The%20Nutraceutical%20Institute.html.

Lakhanpal, T.N. 1986. Mycorrhiza and mushrooms. *In*: Souvenir of mushrooms. National Centre for Mushroom Research and Training, Chambaghat, Solan, pp. 23-25.

Lakhanpal, T.N. 1986. Survey and studies on mushrooms and toadstools of N.W. Himalayas 1983-1986. Final Progress Report DST, GOI Project, New Delhi, pp. 57.

Lakhanpal, T.N. 1988. Morphology and anatomy of ectomycorrhizal fungi from India. *In*: Mycorrhiza Round Table- Being the proceedings of National workshop on mycorrhizae held at JNU, New Delhi March, 13-15, 1987. (Eds. A.K., Verma, A.K., Oka, K.G., Mukerji, K.V.B.R Tilak, and Janak Raj) IDRC, Canada, pp. 53-83.

Lakhanpal, T.N. 1994. Prospects of Mushrooms from the Wild. *In*: Advances in Mushroom Biotechnology (Eds. M.C., Nair, G. Gokulapalan and L. Das) Scientific Publishers, Jodhpur, pp. 15-22.

Lakhanpal, T.N. 1997. Diversity of Mushroom Mycoflora in the North Western Himalayas. *In*: Recent Research in Ecology, Environment and Pollution. Vol. 10 (Eds. S.C. Sati, J. Saxena and R.C. Dubey). Today and Tomorrow's Printers and Publishing House, New Delhi, pp. 35-68.

Lakhanpal, T.N. 2000. Wild and cultivated mushrooms. *In*: Natural Resources and Development in Himalaya. (ed. L.R. Verma). Malhotra Publishing House, New Delhi, pp. 402-417.

Lakhanpal, T.N. and Kaisth, K.R. 1987. Studies on wild edible Mushrooms of Himachal Pradesh (N.W. Himalayas)-III. Ecology and Ethnomycology of two wild edible *Lactarii*. Indian J. Mush. **12-13**: 95-105.

Lakhanpal, T.N. and Shad, O.S. 1986a. Studies on wild edible Mushrooms of Himachal Pradesh (N.W. Himalayas)-I. Ethnomycology, Production and Trade of *Morchella* species. Indian J. Mush. **12-13**: 5-13.

Lakhanpal, T.N. and Shad, O.S. 1986b. Studies on wild edible Mushrooms of Himachal Pradesh (N.W. Himalayas)-II. Ecological relationships of *Morchella* species. Indian J. Mush. **12-13**: 15-20.

Lakhanpal, T.N. and Shad, O.S. 1991. Rhizomorphosphere of morel bearing and non-bearing soil in N.W. Himalayas. *In*: Indian Mushroom. Proceedings of the national Symposium on Mushrooms. (Ed. M.C. Nair), Thiruvananthapuram, Kerala, pp. 172- 178.

Lakhanpal, T.N. and Shad, O.S. 1999. A Re-appraisal of the genus *Morchella* in India. *In*: Advances in Microbial Biotechnology (Eds. J.P. Tewari, T.N. Lakhanpal, Jagjit Singh, Rajni Gupta and B. P. Chamola). A.P.H. Publishing Corporation, New Delhi, pp. 518-523.

Lakhanpal, T.N., Shad, O.S. and Sagar, A. 1991. Mycorrhiza: A possible deterrant in artificial cultivation of morels. Curr. Sci. **60**: 375-377.

Lakhanpal, T.N., Sharma, R. and Singh, L. 1988. Systematics and ectomycorrhizal relationship of some mushrooms in North-Western Himalayas. *In:* Mycorrhizae for Green Asia-being the proceedings of the First Asian Conference on Mycorrhizae, Madras, Jan 29-31, 1988. (Eds. A. Mahadevan, N. Raman and K. Natarajan), pp. 66-69.

Last, F.T., Mason, P.A. and Wilson, J. 1983. Fine roots and the sheathing mycorrhizae: their formation, function and dynamics. Plant and Soil. **71**: 9- 21.

Latif, A., Shinwari, K. Z. and Begum, S. 2004. Potential and Market status of Mushrooms as non-timber forest products in Pakistan. Ethnobotany Project WWF-P, Peshawar, Pakistan.

Lawrence, P. 1952. The association of particular linkage groups with their respective chromosomes in *Neurospora crassa*. Ph.D. Thesis, Columbia University.

Le Roux, P. and Danglot, Y. 1972. Composition chimique et valeur alimentaire du champignon cultive *agaricus bisporus* in function des conditions de stockage. Mush. Sci. **8**: 471-474.

Lee, S.S., Wei, Y.H., Chen, C.H., Wang, S.Y. and Chen, K.Y. 1995. Antitumor effects of *Ganoderma lucidum*. J. Chinese Med. **6**:1-12.

Lee, Y.P. and Takhashi, T. 1966. An improved calorimetric determination of amino acids with use of ninhydrin. Anal Biochem. **14**: 71.

Levisohn, I. 1950. Researchers in soil mycology. Great Britain for Comm. Forest Res. Annu. Report, pp. 83-86.

Lincoff, G.H. 1981. The Audbon Society Field Guide to North American Mushrooms. Simon and Schuster's Publishers, New York, pp. 506.

Litchfield, J.H. 1964. The mass cultivation of *Morchella* species in submerged culture and their potential uses as source of protein. *In*: Global impacts of Allpied Microbiology, M.P. Stars (Ed. John Willey) New York, 327-337.

Litchfield, J.H. 1967. Recent Developments: Submerged culture of morel mushroom mycelium. Food Technology, **21**: 159-161.

Litchfield, J.H., Vely, V.G., and Overbeck, R.C. 1963. Nutrient content of morel mushroom mycelium: Amino acid composition of the protein. J. Food Sci. **14**: 728-741.

Liu, F.A. and Zheng, X. 1993. Study of *Cordyceps sinensis* on anti-laryngeal carcinoma. J. Norman Bethune University Med. Sci. **19**: 57-58.

Liu, K.C., Phounsavan, S.F., Huang, R.L., Liao, C., Hsu, S.Y. and Wang, K.J. 1988. Pharmacological and liver functional studies on the mycelium of *Ganoderma lucidum*. Chinese Pharmaceut. J. **40**: 21-29.

Lombardi, R. and Rountree, R. 2002. Mycological Medicine. Retrieved from the website: file://A:/tnl/mycologicalmedicine.htm.

Lu, B.C. 1962. A new fixative and propionocarmine squash technique for staining fungus nuclei. Can. J. Bot. **40**: 834-847.

Lu, B.C. 1967. The course of meiosis and centriole behaviour during the ascus development of *Gelasinospora calospora*. Chromosoma, **22**: 210-226.

Lunven, P., Le Clement de St. Marcq. C., Carnovale, E., and Fratoni, A. 1972. Nutrition Division of Food and Agriculture Orgenization and National Nutrition Institute, Rome (Unpublished data cited in Food and Agricultural Organization, 1973).

Macris, B.J. and Galiotou-Panayotou, M. 1986. Enhanced cellobiohydrolase production from *Aspergillus ustus* and *Trichoderma harzianum*. Enzyme Microb. Techno. **8**: 141-144.

Maggioni, A. and Renosto, F. 1970. Variazioni della composizione del fungo culthuato (*Agaricus bisporus*) nei process inscatolamento e di conservazione in salamonia. Ind. Conserve. **45**: 311-314.

Maheswari, J.K. 1988. Ethnobotanical research and documentation. Act. Univ. Ups. Symb. Bot. **28**: 207-217.

Mahu, M. 1987. *Morchella vulgaris* Pers. ex Fr. (Morchellaceae) - New from Chile. Mycotaxon, **29**: 373-376.

Mainwaring, H.R. 1967. The role of endoplasmic reticulum in ascomycetes spore delimitation. J. Cell Biol. **35**: 86-87.

Mainwaring, H.R. 1972. The fine structure of ascospores wall formation in *Sordaria fimicola* Arch. Microbiol. **81**: 126-135.

Maire, R. 1905. Recherches cytologiques sur uelques Ascomycetes. Ann. Mycol. **3**: 123-154.

Malloch, D.W. Pirozynski,, K.A. and Raven, P.H. 1980. Ecology and Evolutionary Significance of Mycorrhizal Symbiosis in Vascular Plants (A Review). Proc. Nat. Acad. Sci. (U.S.A.) **77**: 2113-2118.

Manachere, G. 1974. *Coprinus congregatus* un modele bilogique Pour l'etude de quelques Proble'mes generaux poses par la fructification des champignon superieurs. Mush. Sci. **9**: 783-798.

Manachini, P.L., Fortina, M.G. and Parini, C. 1988. Thermostable alkaline protease produced by *Bacillus thermoruber* a new species of *Bacillus*. Appl. Microbiol. Biotechnol. **28**: 409-413.

Marone, D.M., Hendrix, J.W. and Kiernan, J. 1981. Mycorrhizal fungi and their importance in horticultural crop production. New Phytol. **102**: 85-94.

Martin, G.J. 1995, 2005. Ethnobotany: A Method Manual. Chapman & Hall, London.

Marx, D.H. 1969. The influence of ectotrophic mycorrhizal fungi on the resistance of pine roots to pathogenic infections.1. Antagonism of mycorrhizal fungi to root pathogenic fungi and soil bacteria. Phytopathology, **59**: 153-163.

Marx, D.H. 1972.Ectomycorrhizae as biological deterrents to pathogenic root infections. Ann. Rev. Phytopathol. **10**: 429-454.

Marx, D.H. 1976. Use of specific mycorrhizal fungi on tree roots for forestation of disturbed lands. *In*: Proc. Conf. on forestation of disturbed surface areas, Birmingham, Ala. pp. 45-65.

Marx, D.H. 1977. The role of mycorrhizae in forest production. TAPPI Conf. Papers Ann. Meet. Atlanta. G.A. pp. 151-161.

Mateljan, G. 2005. Health Benefits - The World's Healthiest Foods. Retrieved from the website: www.whfoods.com/genpage.php?tname=foodspice &dbid.

Matruchot, L. 1909. Les armillaries. L' armillaie couleur de meil une armillaire digmicose chinoise, Cult. Champignos Comestibles, **25**: 336-338.

Matruchot, L. 1910. Edudes sur la morille. Cult. Champignons Comestible. 121 No. **27**: 418-420, No. **28**: 435-436., No. **29**: 450-451.

Mattila, P., Konbo, K., Eurola, M., Philava, J.M., Astola, J., Vahteristo, L., Hletanlemi, V., Kumpulainen, J., Valtonen, M. and Piironen, V. 2001. Contents of vitamins, minerals elements, and some phenolic compounds in cultivated mushrooms. J. Agric. Food Chem. **49**: 2343-2348.

Mattila, P., Lampi, A. M., Ronkainen, R., Toiva, J. and Piironen, V. 2002 a. Sterol and vitamin D_2 content in some in some wild and cultivated mushrooms. Food Chem. **76**: 293-298.

Mattila, P., Salo-Vaananen, P., Konko, K., Aro, H. Jalava. 2002b. Basic composition of amino acid contents of mushrooms cultivated in Finland. J. Agric. Food Chem. **50:** 6419-6422.

Mayr, R. 1982. Untersuchung zum wachstum und zur Entwicklung eniniger Arten der Mykorrhiza bildenden gutting *Morchella.* Dissertation Universitat Gieben, Germany.

Mc Connell, J.E.W. and Esselen, W.B. Jr. 1947. Carbohydrates in cultivated mushrooms (*Agaricus compestris*) Food Res. **12**: 118-121.

Mc Cubbin, W.A. 1913. The morel. Ontario Nat. Sci. Bull. **8**: 37-40.

McClintok, B. 1945. *Neurospora* – I Preliminary observation of the chromosomes of *Neurospora crassa.* Am. J. Bot. **32**: 671-678.

McCready, R.M., Guqqolz, J., Silviera, Y. and Owns, H.S. 1950. Determination of starch and amylase in vegetables. Ann. Chem. **22**: 1156.

McCubbin, W.A. 1913. The morel – Ontario Nat. Sci. Bull. 8: 37-40.

McKeller, R.L. and Kohrman, R.E. 1975. Amino acid composition of the morel mushroom. J. Agaric. Food Chem. **23**: 464-467.

Mehta, K.B. and Sharma, S.R. 1992. Screening of substrates for mycelial growth and sclerotial formation in *Morchella* species. Indian J. Mycol. Plant Pathol. **22**: 109.

Melin, E. 1922. Boletan-arten als mykorrhizapilze der waldaume. Bor. Deutsch. Bot. Gesellsch. 6: 94-97.

Melzer, U. 1934. Contribution a L'etude Microscopique des Russeles. Bull. Soc. Myc. Fr. **50**: 318-225.

Mendel, L.B. 1989. The chemical composition and nutritive value of some edible American Fungi. Am. J. Physiol. **1**: 225-238.

Mian, G.S. 1988. Himachal Pradesh: History Culture and Economy. Minerva Book House, Shimla, pp. 221.

Milbourne, D., Meyer, R., Bradshaw, J.E., Baird, E., Bonnar, N., Provan, J., Powell, W. and Waugh, R. 1997. Comparison of PCR based marker systems for the analysis of genetic relationships in cultivated potato. Molecular Breeding, **3**: 127-136.

Miles, P.G. and Chang, S.T. 1997. Mushroom Biology–Concise Basis and Current Developments. World Scientific Publishers, pp. 166.

Militoris, H.P. 2001. Mushrooms and Man in Medicine, Myth and Religion. Intern. J. Med. Mush. **3**: 97.

Miller, G.L. 1959. Dinitrosalicylic acid reagent for determination of reducing sugar. Anal. Chem. **31**: 426-428.

Miller, H. 2003. Morel Mushroom Monitoring in a Colorado Burn Area: June-September. Retrieved from the website: http://www .bluewillowpages. com/mushroomexpert/miller_01.html.

Miller, O.K. Jr. 1972. Mushrooms of North America. E.P. Dutton and Company. New York, pp. 360.

Miller, O.K. Jr. 1981. Mushrooms of North America. Elsevier - Dutton Company, New York, pp. 368.

Mizuno, T. 1993. Food Function and Medicinal Effects of Mushroom Fungi. Food Ingred. J. Japan, **22**: 459-464.

Mizuno, T. 1995. Reishi, *Ganoderma lucidum* and *Ganoderma tsugae* bioactive and medicinal effects. Fd. Rev. Intern. **11**: 151-166.

Mizuno, T. 1996. A development of antitumor polysaccharides from mushroom fungi. Fd. & Fd. Ingred. J. **167**: 69-85.

Mizuno, T. 1999. The extraction and development of antitumor active polysaccharides from medicinal mushrooms in Japan (Review). Intern. J. Med. Mush. **1**: 19-30.

Mizuno, T. 2000. The development of an antitumor BRM (Biological Response Modifier) from *Phellinus linteus* (Berk. et Curt.) Teng Mushroom [Review]. Intern. J. Med. Mush. **2**: 21-34.

Mizuno, T. 2002. Medicinal properties and clinical effects of *Agaricus blazei* Murill. Intern. J. Med. Mush. pp. 4.

Mizuno, T. Hagiwara, T., Nakimura, T., Ito, H., Shimura, K., Sumiya, T. and Asakura, A. 1990. Antitumor activity and some properties of water-soluble polysaccharides from "Himematsutake" the fruiting body of *Agaricus blazei* Murill. Agaricult. Boil. Chem.. **54**: 2889-2896.

Mizuno, T., Zhang, J., Kawagishi, H., Nishitoba, T. and Li. J. 1995. Reishi, *Ganoderma lucidum* and *G. tsugae*: Bioactive substances and Medicinal effects. Food Rev. Intern. **11**: 151-166.

Molliard, M. 1904. Mycelium et forme condienne de la morille. Compt. Rend. Acad. Sci. (Paris), **138**: 516-517.

Molliard, M. 1905. Production experimental de L' appeareil ascospore de le morille. Hebd, Compt. Rend. Acad. Sci. (Paris), **140**: 1146-1148.

Montoya, A., Hernandez-Totomoch, O., Estrada-Torres, A., Kong, A. and Caballero, J. 2003. Traditional Knowledge about mushrooms in a Nahua community in the state of Tlaxcala, Mexico. Mycologia, **95**: 793-806.

Morris, B. 1994. Bowa: Ethnomycological notes on the macrofungi of Malawi. *In*: Proceedings of XIII Plenary meeting of AETFAT, Vol. 1 (Eds. J.H. Seyani and A.C. Chikuni). Zomba Malawi, National Herbarium and Botanic Gardens of Malawi, pp. 635-647.

Moser, M. 1949a. Uber des uber das Massenauftreten von Formen der Gattung *Morchella* auf Wanbranflachen. Sydowia, **3**: 174-195.

Moser, M. 1949b. Unterschungen uber den Einflyb von Walbranden auf die Plzvegetation I. Sydowia, **3**: 336.

Moser, M. 1951. Beitrag Zur Anatomie der Discomy. Ceton. Das *Morchella* problem. Sydowia. **5**: 56-119.

Moser, M. 1963. Ascomycetes. *In* Kleine Krptogramenflora (Ed. H. Gams), Innnsbruck.

Munjal, R.L., Sharma, A.D. and Jandaik, C.L. 1977. Edaphic factors in relation to growth of *Morchella* species. Indian J. Mush. **2**: 49-50.

Nakamura, K., Yamaguchi, Y., Kagota, S., Shinozuka, K. and Kunitomo, M. 1999. Activation of Kupffer cell function by oral administration of *Cordyceps sinensis* in rats. Japan J. Pharma. **79**: 505-508.

Nakashima, S., Uneda, T. and Kanada, T. 1979. Effects of polysaccharides from *Ganoderma applanatum* on immune responses. Microbiol. Imunol. **23**: 501-513.

Natarajan, K. 1987. Mycorrhizal fungi associated with *Pinus patula*. *In*: Mycorrhiza Round Table Proc. Natul workshop on Mycrobial, held at JNU (Eds. Verma *et al.*) IDRC, Canada. New Delhi, mare. 13-15. 1987.

Nei, M. and Li, W.H. 1979. Mathematical modeling for studying genetic variation in terms of Restriction Endonuclease. Proc. Natl. Acad. Sci. (USA) **76**: 5269-5273.

Northway, M.C. 2005. Winter 2001 Issue – *Agaricus blazei* mushrooms providing support for a healthy immune system. Retrieved from the website: file://A:/morels/nutraceuticals%20NOW%20winter %202001%20-%*Agaricus*%.

O'Donnell, K.L. 1976. Apothecium morphogenesis in *Peziza quelepidotia*: Scanning electron and light microscopy. Can. J. Bot. **54**: 2254-2267.

O'Donnell, K.L., and Hopper, G.R. 1974. Scanning ultrastructural ontogeny of paragymnohymenial apothecia in the operculate discomycetes *Peziza quelepidotia.* Can. J. Bot. **52**: 873-876.

O'Donnell, K.L., Fields, W.J. and Hooper, C.R. 1970. Apothecial ontogeny in *Saccobolus versicolar* (Pezizales Ascomycetes). Can. J. Bot. **54**: 2055-2060.

O'Donnell, K.L., Flegler, S.L. and Hooper, G.R. 1978. Zygosporangium and Zygospore formation in *Phycomyces nitens*. Can. J. Bot. **56**: 91-100.

Ohno, N., Lino, K., Suzuki, I., Oikawa, S., Sato, K., Miyazaki, T., and Yadomae, T. 1996. Effect of b-glucan on the nitric oxide synthesis of peritoneal macrophage (sic) in mice. Biol. Pharm. Bull. **19**: 608-612.

Olive, L.S. 1965. Nuclear behaviour during meiosis. *In*: The fungi, Vol. I (Eds. G.C. Ainsworth and A.S. Sussman), Academic Press Inc., New York. pp. 119-139.

Olsen, S.R., Cole, C.V., Watanbe, F.S. and Dean, L.A. 1954. Estimation of available phosphorus in solis by extraction with sodium bicarbonate. USDA, Circ.

Ooi, V, E. and Lui, F. 2000. Immunomodulation and anti-cancer activity of polysaccharide - protein complexes. Curr. Med. Chem. 7: 715-729.

Oso, B.A. 1969. Electron microscopy of ascus development in *Ascobolus*. Ann. Bot. **33**: 205-209.

Oso, B.A. 1976. Mushrooms in Yoruba Mythology and Medicinal Practices. Economic Botany, **30**: 367-371.

Overholts, L.O. 1934. The Morels of Pennsylavania. Proc. Penn. Acad. Sci. **8**: 108-114.

Ower, R. 1982. Notes on the development of morel ascocarp: *Morchella esculenta*. Mycologia, **74**:142-44.

Ower, R.D., Mills, G.L. amd Malachowski, J. A. 1986. Cultivation of *Morchella* US Patent No. 4594809.

Ower, R.D., Mills, G.L. amd Malachowski, J. A. 1988. Cultivation of *Morchella* US Patent No. 47557640.

Pacioni, G. 1980. Simon and Schuster's Guide to Mushrooms (Ed. G. Lincoff). Simon and Schuster, New York, pp. 511.

Pal, B.P. 1978. Welcome address i-iv. *In*: Proceedings on National Seminar on Resource Development and environment in Himalayan Region, DST, Govt. of India.

Pathania, M.S. and Vashist, G.D. 2001. Income and Employment Generation from Agribusiness - A Study of Mushroom Farming in Himachal Pradesh. Pol. Economy J. India, **11**: 57-62.

Pegler, D.N. and Fiard, J.P. 1979. Taxonomy and ecology of *Lactarius* (Agaricales) in the Lesser Antilles. Kew Bull. **33**: 601-628.

Phole, P. 1990. Useful Plants of Manang District. A contribution to the Ethnobotany of the Nepal Himalaya, Franz Steiner Veriag Weisboden, GmbH, Stuttgart.

Piearce, G.D. 1985. Livingstone and fungi in tropical Africa. Bull. Brit. Mycol. Soc. **19**: 39-50.

Poope, J. 2000. Use of agricultural waste materials in the cultivation of mushrooms. *In*: Science and Cultivation of Edible Fungi (Ed. D.J.L. Van Grievensen). Rotterdam/Brookfield: A Balkema, pp. 3-24.

Prance, G.T. 1984. The use of edible fungi by Amazonian Indians. *In*: Ethnobotany in Neotropics, (Eds. G.T. Prance and J.A. Kallunki). The New York Botanical Gardens, pp. 127-139.

Purkayastha, R.P. and Chandra, A. 1976. Indian edible mushrooms. Firma Klonput. Ltd. Calcullta, India, pp. 105.

Purkayastha, R.P. and Chandra, A.. 1985. Manual of Indian Edible Mushrooms. Today and Tomorrow's Printers and Publishers, New Delhi, pp. 266.

Qi-Wa lau, 1982. Methods of chemical analysis of mushrooms. *In*: Tropical mushrooms: Biological nature and cultivation methods (Eds. S.T. Chang and T.H. Quimio), The Chinese University Press, Hong Kong, pp. 397-408.

Quackenbush, F.W. Peterson, W.H. and Steenbock, H. 1935. A study of the nutritive value of mushrooms. J. Nutr. **10**: 625-643.

Rai, R.D. 1995. Nutritional and Medicinal values of Mushrooms. *In*: Advances in Horticulture. Vol. 13 (Eds. K.L. Chadha and S.R. Sharma) Malhotra Publishing House, New Delhi. pp. 537-551.

Rai, R.D. and Sohi, H.S. 1988. How protein rich are Mushrooms? Indian Horti. **33**: 2-3.

Rai, R.D., Vijay, B. and Saxena, S. 1990. Production of extracellular enzymes by mesophilic fungi of mushroom compost. Indian Phytopath. **43**: 542-546.

Ramsbottom, J. 1953. Mushrooms and Toadstools: A study of activities of fungi, Collins London, pp. 306.

Ramsbottom, J. 1953. Mushrooms and Toadstools: A study of the activities of fungi, Collins, London, pp. 306.

Rana, M. 2002. Ethnobotanical Studies on the Morels of District Kinnaur (H.P.). M.Phil. Dissertation, Himachal Pradesh University, Shimla-171005.

Ranga Rao, V. and Mukerji, K.G. 1971. Cytology of the ascus development in *Achaetomium strumarium* Rai, Tewari, and Mukerji. Bot. Gaz. **132**: 179-18.

Raper, K.B. and Thom, C. 1949. A manual of the *Penicillia.* Williams and Wilkins Co., Baltimore, pp. 875.

Rathore, M. 2001. Studies on Some Wild Edible Mushrooms of Chamba (Himachal Pradesh). M.Phil. Dissertation, Himachal Pradesh University, Shimla-171005.

Reese, E.T. and Mendel, M. 1963. Enzymatic hydrolase and cellulase and its derivatives. *In*: Methods in carbohydrate chemistry (Ed. R.L. Whistler). (3rd ed.). Academic Press, London, 139-143.

Reeves, F., jr. 1967. The fine structure of ascus formation in *Pyronema domesticum.* Mycologia, **59**: 1018-1033.

Reilly, C. 1996. Selenium in food and health. Blackie academic & Professional, London.

Repin, C. 1901. Sur la culture de la morille. Rev. gen. des. sciences, **12**: 595-596.

Rholf. F.J. 1998. NTSYS - pc. Numerical Taxonomy and Multivariate Analysis System, Version 2.0. Exeter Software, New York.

Rifai, M.A. 1968. The Australian Pezizales in Herbarium of the Royal Botanical Gardens Kew. Verh. K. Ned. Akad. Wet. Afd. Natuurkd. Reeks 1, **57**: 1-295.

Robbins, J.W. and Hervey, A. 1965. Manganese, Calcium filtrate factor for *Morchella crassipes.* Mycologia, **57**: 262-274.

Robbins, W.J. and Hervey, A. 1969. Wood extract and growth of *Morchella.* Mycologia, **51**: 356-363.

Robert, E. 1865. Relations centre la famille des Olenees et les morilles. Bulletin de Pa Souete botanique de France, **12**: 224-246.

Robert, E. 1865. Relations entre la famille desoleineas et les moriller.Bulletin de la Societe potanique de France. **12**: 244-246.

Robinson, R.F. and Davidson, R.S. 1959. The large-scale growth of higher fungi. Appl. Microbiol. **1**: 216-278.

Rogers, J.D. 1965. *Hypoxylon fuscum* I - Cytology of the ascus. Mycologia, 57: 789-803.

Rojas, C. and Mansur, E. 1995. Ecuador: informaciones generales sobre productos non madereros en Ecuador. *In:* Memoria, consulta de expertos sobre productos forestales no madereros para America Latina y el Caribe. Santigo, Chile, FAO Regional Office for Latin America and the Caribbean, pp. 208-223.

Rolfe, R.T. and Rolfe, F.W. 1925. The Romance of the Fungus World. Chapman and Hall Ltd., London.

Romagnesi, H. 1963. Petit atlas des Champignos 3. Paris, Bordas, pp. 286.

Royse, D. J. 1995. Speciality Mushrooms - Cultivation on synthetic substrate in the USA and Japan. Mush.

Royse, D.J. and May, B. 1990. Interspecific allozymes variation among *Morchella* spp. and its inference for systematics within the genus. Biochem. Syst. Ecol. **18**: 475-479.

Roze, M.E. 1883. La parasitisme du *Morchella esculenta* Pers. Sur I' *Helianthus tuberous* L. Bull. Soc. Bot. Franc. **30**: 139-143.

Saccardo, P. A. 1889. *Sylloge Fungorum huscusque Congitorum,* Patavii, pp. 8.

Saccardo, P.A. 1882-1931. Sylloge Fungorum. VIII 1889.

Saini, S.S. and Atri, N.S. 1999. Exploring mushroom biodiversity for pharmaceutical activity. *In*: From Ethnomycology to Fungal Biotechnology: Exploiting Fungi from Natural Resources for Novel Products (Eds. J. Singh and K.R. Aneja). Kluwer Academic/Plenum Publishers, New York, pp. 41-49.

Samajpati, N. 1978. Nutritive value of some Indian edible mushrooms. Mush. Sci. **10**: 695-703.

Samorini, G. 1992. The oldest representations of hallucinogenic mushrooms in the world (Sahara Desert, 9000-7000 B.P.). Integration, **2**: 69-78.

Samorini, G. 1994. La religion Buiti y la planta psicoactiva Tabernanthe ibogo, Africa Ecuatorial. *In*: Plants, Chamanismoy Estados de Consciencia (Ed. J. Fericgla). Barcelona, Los Libros de la Liebre de Marzo, pp. 175-195.

Samorini, G. 1995a. Gli Allucinogenic nel Mitol Raccontisulle origini delle pinte psicosttive, Torine, Nautilus.

Samorini, G. 1995b. Umbrella-stones or Mushroom-stones? (Kerala, southern India). Integration, **6**: 33-40.

Samorini, G. 1995c Traditional use of psychoactive mushrooms in Ivory Coast. Eleusis, J. Phychoactive Plants Comp. **1**: 2-27.

Samorini, G. 2001. New data from the Ethnomycology of Psychoactive mushrooms. Intern. J. Med. Mush. **3**: 257-278.

Santanna, E.S., Teixeira, E. and Morette, E. 1986. *Morchella crassipes* growth in synthetic and vinase media. Rev. Microbiol. **17**: 10-14.

Saqib, Z. and Sultan, A. 2004. Ethnobotany of Palas Valley. Retrieved from website: www.siu.edu/~ebl/leaflets/zafeer.htm.

Scherer, J. 1979. High altitude ethnobotany of Rolwaling Sherpas. Contributions to Nepalese Studies, **12**: 45-64.

Schmidt, E.L. 1979. Puffing in *Morchella*. Bull. Brit. Mycol. Soc. **13**: 126-127.

Schmidt, E.L. 1983. Spore germination of and carbohydrate colonization by *Morchella esculenta* at different soil temperatures. Mycologia, **75**: 870-875.

Schultes, R.E. 1962. The role of ethnobotanist in the research for new medicinal plants. Lloydia, **25**: 257-266.

Schultes, R.E. 1940. Teonancatl: the narcotic mushroom of the Aztecs. American Anthropologist, **40**: 429-443.

Seaver, F.J. 1928. The North American cup fungi (Operculates). Seaver. New York, pp. 284.

Seaver, F.J. 1942. The North American cup fungi (Operculates). Hafner Publishing Co., New York, pp. 370.

Seekopf, L. and Schuster, H. 1957. Qualitative and Quantitative Aminosaurebestimmungen an eingen wichtigen Speisepilzen. Z. lebensm. Unters. Forsch. **106**: 177-187.

Sehgal, S. and Sharma, B.M. 2007. Studies on heterokaryon formation in *Morchella esculenta*. Int. Conf. on Mushroom Biology and Biotechnology. NRCM, Solan, 10-11 Feb. 2007, p. 47 (Abstr.)

Shad, O.S. 1985. Studies on some edible fungi of Himachal Pradesh. M.Phil. Dissertation, Himachal Pradesh University, Shimla–171 005.

Shad, O.S. 1989. Biological studies on *Morchella* species (morels) of Himachal Himalayas. Ph.D. Thesis, Himachal Pradesh University, Shimla–171 005.

Shad, O.S. and Lakhanpal, T.N. 1991. Sociobiology of morels in N.W. Himalayas. *In*: Indian Mushroom. Proceedings of the national Symposium on Mushrooms. (Ed. M. C. Nair). Thiruvananthapuram, Kerala, pp. 1-3.

Shad, O.S., Sagar, A. and Lakhanpal, T.N. 1990. The mycorrhizal association of morels in N.W. Himalayas. *In:* Current Trends in Mycorrhizal Research. (Eds. B.L. Jalali and S. Chand). HAU, Hissar, pp. 128-129.

Sharma, B. 1987. Studies on Non-Gilled Mushrooms of Himachal Pradesh. M.Phil. Dissertation, Himachal Pradesh University, Shimla-171 005.

Sharma, B. 1993. Studies on Some Wild Edible Fungi of Himachal Pradesh. Ph.D. Thesis, Himachal Pradesh University, Shimla-171 005.

Sharma, J.R. 1980. Systematics and ecology of family Boletaceae. Ph.D. Thesis, Himachal Pradesh University, Shimla-171005.

Sharma, J.R. and Lakhanpal, T.N. 1981. Mycorrhiza forming species of Boletaceae. *In*: Proc. Symposium on forest biomass (Ed. P.K. Khosla). Indian Soc. of Tree Scientists, Solan, H.P. pp. 455-457.

Sharma, M. 2001. Studies on Some Wild Edible Mushrooms of Hamirpur (Himachal Pradesh). M.Phil. Dissertation, Himachal Pradesh University, Shimla-171 005.

Sharma, R. 1986. Studies on the family Boletaceae of Himachal Pradesh, M.Phil. Dissertation, H.P. University, Shimla, pp. 145.

Sharma, S. 2002. Physiological and Molecular Characterization of edible ectomycorrhizal fungus *Cantharellus cibarius*. M.Sc. Dissertation, Department of Biotechnology and Environmental Sciences, TIET, Patiala - 147 004.

Shimizu, A., Yano, T., Saito, Y. and Inada, Y. 1985. Isolation of an inhibitor of platelet aggregation from a fungus *Ganoderma lucidum*. Chem. Pharm. Bull. **33**: 3012-3015.

Shrestha, P. 1988. Contributuion to the ethnobotany of the Tamangs of the Kathmandu Valley. Contributions to Nepalese Studies, **15**: 247-266.

Sillitoe, P. 1995. Ethnoscientific observations on entomology and mycology in the southern highlands of Papua New Guinea. Science in New Guinea, **21**: 3-26.

Singer, R. 1961. Mushrooms and Truffles. Leonard Hill London, New York, pp. 272.

Singer, R. and Harris, B. 1987. Mushrooms and Truffles. Scientific Books, Germany, pp. 389.

Singer, R. and Morello, J.H. 1960. Ectotrophic mycorrhiza and Forest Communities. Ecology, **41**: 549-551.

Singh, J. 1999. Ethnomycology and Folk Remedies: Fact and Fiction. *In*: From Ethnomycology to Fungal Biotechnology: Exploiting fungi from Natural Resources for Novel Products (Eds. J. Singh and K.R. Aneja). Kluwer Academic/Plenum Publishers, New York, pp. 11-17.

Singh, L. 1988. Studies on an edible mycorrhizal hypogeous fungus. M.Phil. dissertation, Himachal Pradesh University, Shimla-171005.

Singh, S.C. 1996. Mushrooms. Deorali, **4**: 15-31.

Singh, S.K., Kamal, S., Tiwari, M., Yadav, M.C. and Upadhyay, R.C. 2004. Arbitrary primer based RAPD: A useful genetic marker for species identification in morels. J. Plant Biochem. Biotechnol. **13**: 7-12.

Singh, S.K., Rai, R.D., Rai, A.K. and Verma, R.N. 2001. Production of extracellular lignolytic enzymes by *Morchella esculenta*. Mush. Res. **10**: 99-102.

Singh, S.K., Verma, R.N. and Kamal, S. 2003. DNA fingerprinting of edible mushrooms using Random primers. *In*: Current Vistas in Mushroom Biology and Production (Eds. R.C. Upadhay, S.K. Singh and R.D. Rai). Mushroom Society of India, pp. 57-60.

Singleton, J.R. 1953. Chromosome morphology and the chromosome cycle in the ascus of *Neurospora carassa*. Am. J. Bot. **40**: 124-144.

Smith, A.H. and Weber, N.S. 1980. The Mushroom Hunter's Field Guide. University of Michigan Press, Ann. Arbor. pp. 316.

Smith, A.H., Smith , H.V. and Weber, N.S. 1981. How to know the non-gilled mushrooms. Williams C. Brown Co. Publishers, Dubuque, Iowa, U.S.A., pp. 316.

Smith, A.H., Smith H.V. and Weber, N.S. 1981. How to know the non-gilled mushrooms. William C. and Brown Co. Publishers, Dubuque, Iowa, U.S.A.

Smith, E.J., Rowan, N.J. and Van Kok-Kheng. 2000. Functional food science and the medicinal mushroom. Intern. J. Med. Mush. **2**: 277-28.

Smith, J., Rowan, N. and Sullivan, R. 2002. Medicinal Mushrooms: Their therapeutic properties and current medical usage with special emphasis on cancer treatments. University of Strathelyde, Glasgow, UK, pp. 256.

Smith, J.E. and Berry, D.R. 1974. Biochemistry of fungal development. Academic Press, London.

Sohi, H.S., Kumar, S. and Seth, P.K. 1965. Some interesting fleshy fungi from Himachal Pradesh. J. Indian Bot. Soc. **54**: 69-73.

Sorensen, T. 1948. A method of establishing groups of equal amplitude in plant sociology on similarity of species content. Det Kong. Danska Vedensk. Selsk-Biol. Skr. (Copenhagen) **5**: 1-34.

Spaulding, T. and Beelman, R.B. 2003. Survey evalution of selenium and other minerals in *Agaricus* mushroom commercially grown in United States. Mush. News, **51**: 6-9.

Spolar, M.R. Schaffer, E.M. Beelman, R.B. and Milner, J.A. 1999. Selenium-enriched *Agaricus bisporus* mushrooms suppress 7, 12-dimethylbenz (A) anthracene bioactivation in mammary tissues. Cancer Letters, **138**: 145-150.

Stamets, P. 2000. Growing Gourmet and Medicinal Mushrooms. Ten Speed Press, Berkeley.

Steirs, D.L. 1974. Fine structure of ascospore formation in *Poronia punctata*. Can. J. Bot. **32**: 999-1003.

Stingve, T. 1977. Selenium content of mushrooms. Z Lebensm Unters Forsch. **164**: 201-203.

Subedi, B.P. Plant profile. 2004. Retrieved from the website:www.ansab.org News_Fol/newsletter-Vol5/Plant_profile.htm.

Susuki, S. and Oshima, S. 1974. Influence of shii-ta-ke, *Lentinus edodes*, on human serum cholestrol. Mush. Sci. **9**: 463-467.

Sydow, H. and Butler, E.J. 1911. Fungi Indiae Orientalis part III. Ann. Mycol. **9**: 372-421.

Sydow, H. and Buttler, E.J. 1911. Fungi Indiae Orientalis part III. Ann. Mycol. **9**: 372-421.

Taylor, J. and Natvig, D.O. 1987. Isolation of fungal DNA. *In*: Zoosporic Fungi in Teaching and Research (Eds. M. S. Fuller and A. Jaworski). Athens, Georgia, Southeastern Publishing Co-operation, pp. 252-258.

Thakur, K. 1990. Mushroom Recepies–Indian Blending. *In*: Recent developments in the cultivation technology of edible mushrooms. (Ed. K.B. Mehta) National Centre for Mushroom Research and Training, Chambaghat, Solan, pp. 143-147.

Theissen, F. 1911. Fungi aliquot Bombyenses a Rev. Dd. Blatter collecti. Ann. Mycol. **9**: 153-159.

Theissen, F. 1911. Fungi aliquot Bombyenses a Rev. Ed. Blatter Collecti. Ann. Mycol. **9**:153-59.

Thomson, T. 1847-1848. Western Himalayas and Tibet, a narrative of a journey through the mountains of Northern India. London.

Tilak, T. and Rao, R. 1968. Second Supplement to the fungi of India. Aurangabad Printers. Aurangabad, pp. 312.

Tiwari, M., Singh, S.K., Kamal, S., Singh, S. and Yadav, M.C. 2004. Extracellular enzyme polymorphism in *Morchella* and related species. Mush. Res. 98: 114-118.

Trappe, J.M. 1971. Mycorrhiza forming Ascomycetes. *In*: Mycorhizae (Ed. Hacskaylo, U.S.D.A.) For Serv. Misc. Publ. 1189, pp. 19-37.

Tuite, J. 1960. Plant Pathological Methods: Fungi and Bacteria. Burges Publishing Co. Minn. U.S.A., pp. 239.

Tuno. 2001. Mushroom utilization by the Majangir, an Ethopian tribe. Mycologist, **15**: 78-79.

Turner, N.J., Kuhnlein, H.V. and Egger, K.N. 1986. The Cotton Wood Mushroom (*Tricholoma populinum*), a food resource of the Interior Salish Indian people of British Columbia. Can. J. Bot. **65**: 921-927.

USDA, 2001. USDA Nutrient Database for Standard Reference. Release 14, Nutrient Data Laboratory, Retrieved from the website: http://www.nal.usda.gov/fnic/cgi-bin/nut_search.pl.

Vaidya, J.G. and Rabba, A.S. 1993. Fungi in Folk Medicine. Mycologist, **7**: 162-166.

Veilux, C. and Steven, R. K. 1996. An Introduction to Ethnobotany. Shainan Pharmaceuticals, Inc. Work interdisciplinary to discover new drugs. (Ed. L. Morganstein), pp. 1-6.

Villarreal, L. and Perez –Moreno, J. 1989. Los hongos comestibles silvestres de Mexico, un enfoque integral. Micologia Neotropica Aplicada, **2**: 77-114.

Volk, T.J. 1990. The current state of morel cultivation. Mushroom News, 1999: 24-27.

Volk, T.J. and Leonard, T.J. 1989a. Physiology and environmental studies on sclerotial formation and maturation in *Morchella.* Appl. Env. Microbiol. **55**: 3094-3100.

Volk, T.J. and Leonard, T.J. 1989b. Experimental studies on the morel I. Heterokaryon formation between monoascosporus strains of *Morchella.* Mycologia, **81**: 523-531.

Volk, T.J. and Leonard, T.J. 1990. Cytology of life cycle of *Morchella.* Mycol. Res. **94**: 399-406.

Von, M. 1861-1871. Journeys and Investigations in the Jakutskaia Oblast of Eastern Siberia in 1861-1871.

Vos, P., Hogers, R., Bleeker, M., Reijans, M., Van Der Lee, T., Hornes, M., Frijters, A., Pot, J., Peleman, J., Kuiper, M. and Zabeau, M. 1995. AFLP: A new concept for DNA fingerprinting. Nucl. Acids Res. **21**: 1612-1645.

Wakode, D.D. 1983. Distribution of *Morchella* in time and space with special reference to Melghat forests. *In*: Silver Jubilee Symposium on 'Science and Cultivation of Edible Fungi'. Regional Research Laboratory, Sanatnagar, Srinagar, pp. 87 (abst.).

Walters, B. 1995. Halleluja. Praise the mushrooms. Psyched. III.

Wang, H. X., Liu, W.K., Ng, T.B., Ooi, V.E.C. and Chang, S. T. 1995. Immunomodulatory and antitumor activities of a poly-peptide complex from a mycelial culture of *Trichoderma* sp., a local edible mushroom. Life Sci. 57: 269-281.

Wang, R., Xu, Y., Ji, P., Wang, X. and Holliday, J. 2001. Clinical trials of a mixture of six medicinal mushrooms extracts. Retrieved from the website: http://alohamedicinals.com/clinicals_trials.html.

Wang, Y.C. 1987. Mycology in ancient China. Mycologist, **1**: 59-61.

Waraitch, K.S. 1976. The genus *Morchella* in India. Kavaka, **4**: 69-76.

Warcup, J.H. 1960. The ecology of soil fungi. *In:* (Ed. D. Parkinson and J.S. Waid) Liverpool, University Press, pp. 3-21.

Wasser, S. P. 2000. Medicinal mushrooms as a source of antitumor and immunomodulatory polysaccharides. Retrieved from the website: http://Limk.springer.de/link/service/ journals/00253/contents/02/01076/index.html.

Wasser, S.P. 2002. Review of Medicinal Mushroom Advances: Good News from Old Allies. Herbal Gram, **56**: 28-33.

Wasser, S.P. and Weis, A.L. 1997. Medicinal mushrooms. *In*: Reishi Mushroom (*Ganoderma lucidum* (Curt.: Fr.) P. Karst.) (Ed. E. Nevo). Haifa, Israel: Peledfus, pp. 1-37.

Wasser, S.P. and Weis, A.L. 1999. Medicinal properties of substances occurring in Higher Basidiomycete mushrooms: current perspectives. Intern. J. Med. Mush. **1**: 31-62.

Wasser, S.P., Nevo, E., Sokolov, D., Reshetnikov, S. and Timor – Tismenetsky, M. 2000. Dietary supplements from medicinal mushrooms: diversity of types and variety of regulations. Intern. J. Med. Mush. **2**: 1-19.

Wasson, R.G. 1968. Soma: Divine Mushroom of Immortality. The Hague, Mouton, pp. 381.

Wasson, R.G. 1980. The wondrous mushrooms. McGraw-Hill Book Co., New York, pp. 248.

Wasson, V.P. and Wasson, R.G. 1957. Mushrooms, Russia and History. Vol. 1 & 2, Pantheon, New York.

Weber, N.S. 1988. A morel hunter's companion. Two Penninsula Press, Lansing, Michigan, pp. 111-167.

Weber, N.S. and Smith, A.H. 1985. A field guide to Southern Mushrooms. Univ. of Michigan Press, Ann. Arbor. pp. 280.

Webster, J. 1970. Introduction to Fungi. Cambridge University Press, Cambridge.

Wells, K. 1970. Light and electron microscopic studies of *Ascobolus stercorarius.* I. Nuclear divisions in the ascus. Mycologia, **62**: 761-790.

Wells, K. 1972. Light and electron microscopic studies on *Ascobolus stercorarius.* II. Ascus and ascospore ontogeny. Uni. Cal. Publ. Bot. **62**: 1-93.

Werner, A.R. and Beelman, R.B. 2001. Growing Selenium-enriched mushrooms as ingredients for functional foods or dietary supplements. Intern. J. Med. Mush. **3**: 247.

Wiletts, H.J. and Bullock, S. 1992. Developmental Biology of sclerotia. Mycol. Res. **96**: 801-816.

Wiletts, H.J., 1978. Sclerotium formation in the filamentous fungi 3. *In*: Developmental Mycology (Eds. J.E. Smith and D.R. Berry) Edward Arnold, London, pp. 197.

Wilhelm, S. 1966. Chemical treatment and inoculum potential of soils. Ann Rev. Phytopathol. 4: 53-78.

Wilkins, W.H. and Harris, G.C.M. 1946. The ecology of larger fungi - V. An investigation into the influence of rainfall and temperature on the seasonal production of fungi in beech-wood and pine wood. Ann. Appl. Biol. **33**: 179-188

Wilkins, W.H. and Patrick, S.H.M. 1939. The ecology of larger fungi–III. The constancy and frequency of grassland types with special reference to the influence of environmental factors. Ann. Appl. Biol. **26**: 25-46.

Wilkins, W.H. and Patrick, S.H.M. 1940. The ecology of larger fungi–IV. The seasonal frequency of grassland fungi with special reference to the influence of environmental factors. Ann. Appl. Biol. **27**: 17-34.

Wilkins, W.H., Ellis, E.M., and Harley, J.L. 1937. The ecology of larger fungi–I. The constancy and frequency of fungal species in relation to certain vegetation communities, particularly oak and beech. Ann. Appl. Biol. **24**: 703-732.

Wilkins, W.H., Harley, J.L. and Kent, S.C. 1938. The ecology of larger fungi–II. The distribution of larger fungi in part of Charlton forest. Ann. Appl. Biol. **25**: 472-489.

Williams, J.G.K. and Kubelik, A.R. 1991. Genetic analysis with RAPD markers. *In*: More Manipulations in fungi (Eds. J.W. Bennett and L.L. Lasure). Academic Press, New York, pp. 431-439.

Williams, J.G.K., Kubelik, A.R., Livak, K.J., Rafalski, J.A. and Tingey, S.V. 1990. DNA polymorphism amplified by arbitrary primers are useful as genetic markers. Nucl. Acids Res. **18**: 7213-7218.

Wipf, D., Bedell, J.P., Munch, J.C., Botton, B. and Buscot, F. 1996. Polymorphism in morels: Isozyme electrophoretic analysis. Canadian J. Microbiol. **42**: 819-827.

Wipf, D., Munch, J.C., Botton, B. and Buscot, F. 1996. DNA polymorphism in morels: Complete sequences of the Internal Transcribed spacer of genes coding for rRNA in *Morchella esculenta* (yellow morels) and *M. conica* (Black morels). Applied and Environ. Microbiol. **62**: 3541-3543.

Wooster, H.A. Jr. 1954. Nutritional data, H.J. Heinz Co., Pittsburgh.

Yamura, Y. and Cochran, K. W. 1976. Chronic hypocholesterolemic effect of *Lentinus edodes* in mice and absence of effect on scarpie. Mush. Sci. **9**: 489-493.

Yoon, C., Gessner, R.V. and Romano, M.A. 1990. Population genetics and systematics of *Morchella esculenta* complex. Mycologia, **82**: 227-235.

Young, H.E. 1936. A mycorrhiza forming fungus. *Pinus.* J. Aust. Inst. Agri. Sci. **2**: 32-34.

Young, H.E. 1940. Mycorrhizae and growth of *Pinus* and *Araucaria.* The influence of different species of mycorrhiza forming fungi on seedling growth. J. Aust. Inst. Agri. Sci. **6**: 21-25.

Yuan, C.Z., Hei, P.M., Cui, J.H. and Takeuchi, H. 1998. Hypoglycemic effect of water soluble polysaccharide from *Auricularia auricula-judae* Quel. on genetic diabetic mice. Biosci. Biotech. and Biochem. **62**: 1898-1903.

Zak, B. 1971. Characterization and identification of Douglas fir mycorrhizae. *In:* Ectomycorrhize – Ecology and Physiology (Eds. G.C. Marks and T.T. Kozlowski), Academic press, New York, pp. 43-78.

Zakany, J., Chihara, G. and Fachet, J. 1980a. Effect of Lentinan on tumour growth in Murine allogenic and syngeneic host. Intern. J. Cancer, **25**: 371-376.

Zakany, J., Chihara, G. and Fachet, J. 1980b. Effect of Lentinan on the production of migration inhibitory factor induced by syngeneic tumour in mice. Intern. J. Cancer, **26**: 783-788.

Zang, M. 1987. Some new and noteworthy higher fungi from eastern Himalayan (China) Acta Bot. Yunnanica, **9**: 81-88.

Zhang, Z. and Xia, S.S. 1990. *Cordyceps sinensis* - 1 as an immunosuppressant in hetero suppressant in heterotrophic heart allograft in rats. J. Tougji. Medicine. Univ. **10**: 100-103.

Zhu, H., Jia-Shi, Halpern, G. and Jones, K. 1998. The Scientific Rediscovery of an Ancient Chinese Herbal Medicine: *Cordyceps sinensis*. Journal of Alternative Complementary Medicine, **4**: 289-303.

Zhuang, C. and Mizuno, T. 1999. Biological responses from *Grifola frondosa* (Dick.: Fr.) S. F. Gray - Maitake (Aphyllophoromycetideae). Intern. J. Med. Mush. **1**: 317-324.

Zhuang, C., Mizuno, T., Ito, H., Shimura, K., Sumiya, T. and Kawade. 1994a. Fractionation and anti-tumour activity of polysaccharides of *Grifola frondosa.* Biosci. Biotechol. Biochem. **58**: 185-188.

Zhuang, C., Mizuno, T., Ito, H., Shimura, K., Sumiya, T. and Kawade, M. 1994b. Anti-tumour activity and immunological property of polysaccharides from the mycelium of cultured *Grifola frondosa* mycelium. Nippon Shokuhin Kogyo Gakkaishi, **41**: 724-732.

Zietkiewicz, E., Rafalski, A. and Labuda, D. 1994. Genome fingerprinting by simple sequence repeat (SSR)–anchored polymerase chain reaction amplification. Genomics, **20**: 176-183.

Index